THINKING KENKEN

STRETCH YOUR BRAIN AND ENTERTAIN!

150 Puzzles That Make You Smarter

Written by:
Noah Tirschwell

Created by:
Tetsuya Miyamoto

KENKEN
www.kenkenpuzzle.com

To access free, unlimited puzzles of all sizes and difficulty levels, visit www.kenkenpuzzle.com

Download the FREE KenKen Classic app for iOS and Android, available in the App Store and Google Play.

Thinking KenKen: Stretch Your Brain and Entertain!

www.kenkenpuzzle.com

Project editors: Tyler Kennedy, Jerry March
Designer: Tyler Kennedy

ISBN-13: 978-1-945542-11-4
First Edition: April 2021

THINKING KENKEN

STRETCH YOUR BRAIN AND ENTERTAIN!

Noah Tirschwell

When I was given my first KenKen at age seven back in 2011, I could not wrap my head around the odd-looking grid of numbers in front of me. Within minutes, I was so upset and confused by the seemingly herculean task of solving these puzzles that I found myself crying over — you guessed it — a mere 3x3.

And so began my journey with KenKen. Over the years, we have built quite a relationship. I participated in the annual grade-wide KenKen competition at my elementary school for four years, during which I met the puzzle's brilliant creator, Tetsuya Miyamoto. In the past six years, I've been to five International KenKen Competitions, each one an incredible experience. I've watched KenKen World Champion John Gilling tackle beastly puzzles with total ease. I've been introduced to Will Shortz, the New York Times Puzzle Editor and world's most respected puzzle guru, and watched him play a wicked game of ping-pong. I've met teenagers from all around the world. And I've even been able to move up the scoreboard to the point that I am currently the two-time defending high school champion!

But doing KenKen is not solely about the competition — that much has always been clear to me. KenKen is about relaxation. It is calming and stress relieving (except, of course, for the puzzles that are so stressful that you eventually tear them up in utter frustration)! And above all else, this puzzle is enjoyable. It doesn't matter if I'm completing a quick 4x4 on my phone in the car or sitting by a fire in the dead of winter inching my way through a thick book of 9x9s. Regardless of where I am or how I am feeling, I can immerse myself in KenKen and block out any background noise.

And that is what makes KenKen so irresistible. It allows you to let go of your surroundings and simultaneously demands your utmost focus and perseverance. KenKen combines logic, numbers, and basic arithmetic to create a world that is stimulating yet calming, challenging yet enjoyable. This is the magic of KenKen.

Some KenKen puzzles can be completed in seconds without any serious thought. Others are frustratingly difficult and can require over a half-hour of intense and uninterrupted attention. However, despite this far ranging spectrum, there exists a balance between two extremes. These are the KenKens that make you pause for a minute while you search for a breakthrough. But eventually, a path to the solution always appears. These KenKens require focus and attention, and they undoubtedly make you think. Yet you relax while you do them, and working through them is just a total joy.

So when I set out writing this book, I wanted to generate these kinds of puzzles. KenKens that are stimulating yet calming, challenging yet enjoyable. I sought to put together a batch that was neither too easy nor too difficult, neither stressless nor stressful. And above all else, they needed to make you think.

As you make your way through this book, you'll find a compilation of 150 "Thinking KenKen." Some will inevitably be easier than others, but each puzzle was designed with the same intent. These are KenKens that will make you think, KenKens that require you to think, KenKens that require logic, KenKens that will make your day. So kick back, relax, start thinking, and let's KenKen!

ACKNOWLEDGEMENTS

I am very grateful to everyone who made this book possible. First, a shout-out to my incredible elementary school math specialist, Mr. Josh Rosen, who first introduced me to KenKen, and whose contagious passion cultivated in me a deep love for math puzzles. Thank you to my family — Oren, Talia, Mom and Dad — for their support in all my KenKen endeavors. Thank you to Jerry March, who helped me through various stages of the book process and to Tyler Kennedy, who did a fantastic job both designing the cover and piecing together the many parts of the book. The biggest possible thank you goes to Bob Fuhrer, who gave me the opportunity to create this KenKen book and patiently guided me at every step of the process. And finally, thank you to the genius behind KenKen: Tetsuya Miyamoto. His brilliant contributions to the math puzzle world have changed my life and I am deeply grateful to him.

ABOUT THE AUTHOR

Noah Tirschwell is a junior at Hackley High School and lives in Dobbs Ferry, New York with his brother, sister, mom and dad. He has been an avid math puzzler since a very young age, and currently holds the title of two-time International KenKen Champion for the high school (Sigma) division. Outside of the KenKen world, Noah is a serious violinist and is passionate about doing social justice work in his community.

How To Solve

◊ Fill each square in the grid with a single number, using all numbers up to the puzzle size (ex. a 5x5 puzzle uses numbers 1 through 5).

◊ Do not repeat a number in any row or column.

◊ Each heavily outlined set of squares is called a "Cage".

◊ The numbers in each Cage must combine – in any order – to produce the Target Number indicated in the top corner of the Cage by using the math operation next to the Target Number. For example, in a 6x6 puzzle, if the Target Number in a 2-square Cage shows "3-", the numbers in that cage must total 3 in any order, by using subtraction (4-1, 5-2, 6-3).

◊ A number may be repeated within a 3-square or larger Cage as long as it is not in the same row or column. You can do this when a Cage has an "L" shape or spans several rows or columns.

Hint: To start, fill in the single-square Cages first. Since no calculation is required and the answer is "given" to you, we call these "Freebies."

And that's it!

Solving a KenKen puzzle involves pure logic and mathematics. No guesswork will ever be needed, and each puzzle has only one solution. So sharpen your pencil, put on your thinking cap, and find out why Will Shortz, NPR Puzzle Master and The New York Times Puzzle Editor calls KenKen "The most addictive puzzle since Sudoku!"

Visit www.kenkenpuzzle.com any time to play unlimited puzzles of all sizes and difficulty levels, absolutely free of charge.

1

+ − × ÷

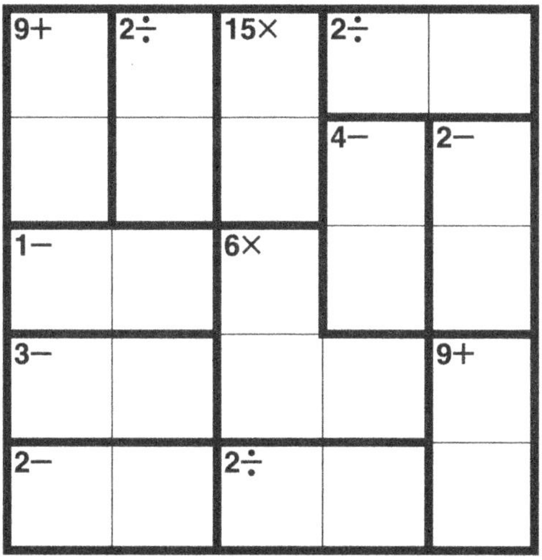

2

+ − × ÷

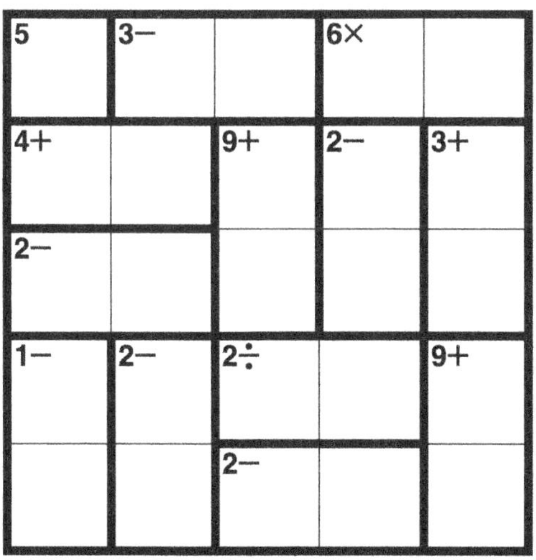

3 + − × ÷

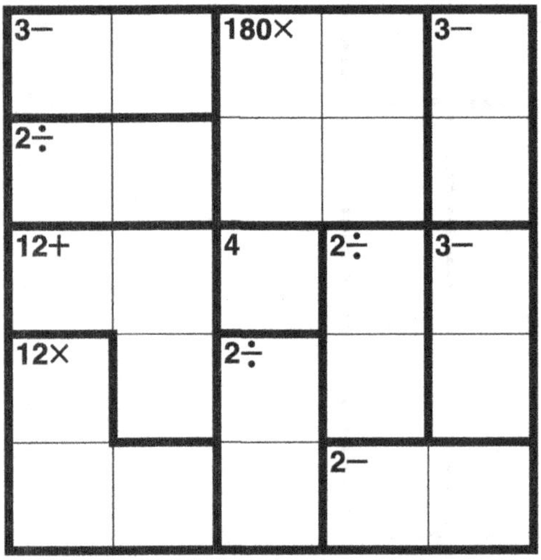

4 + − × ÷

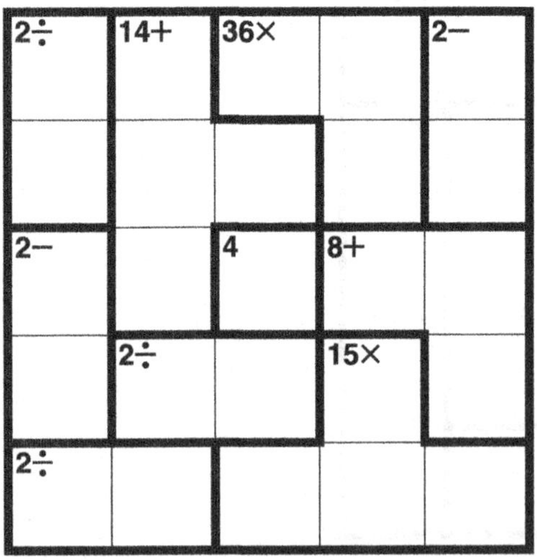

5

+ − × ÷

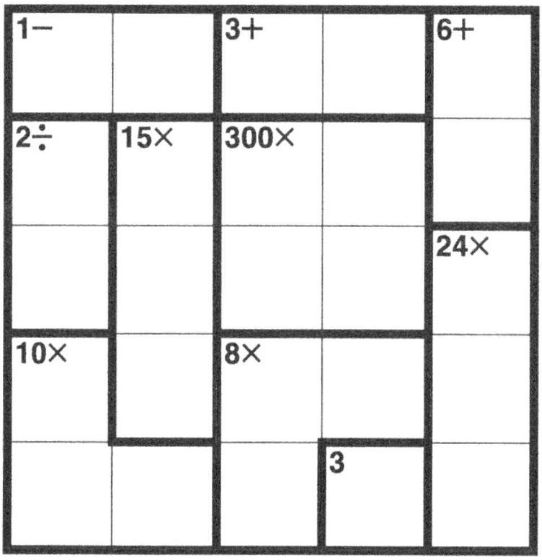

6

+ − × ÷

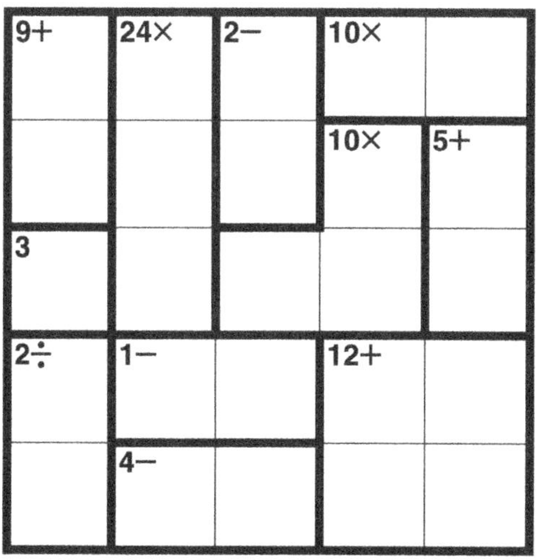

7 + − × ÷

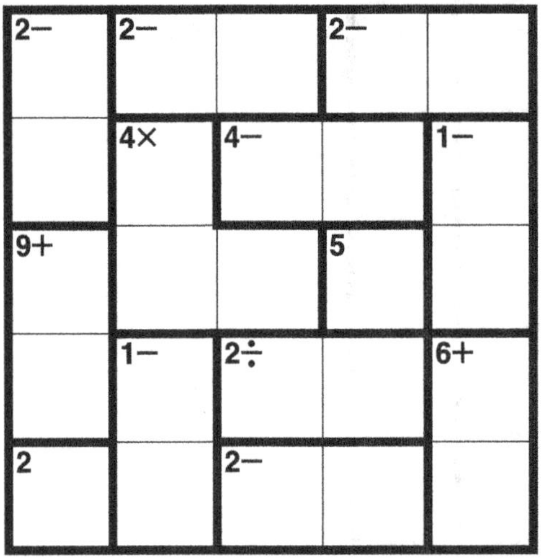

8 + − × ÷

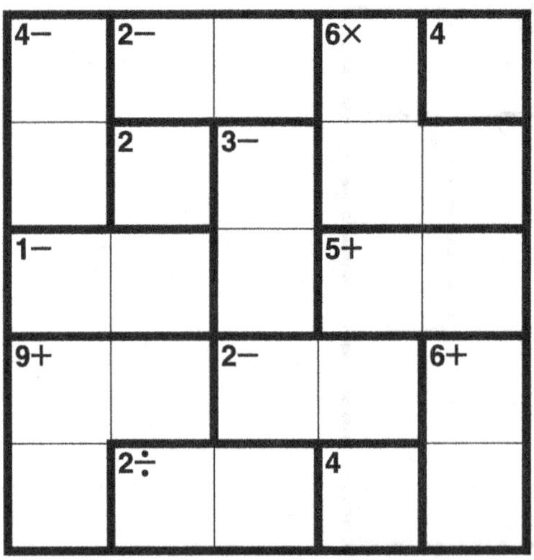

9 + − × ÷

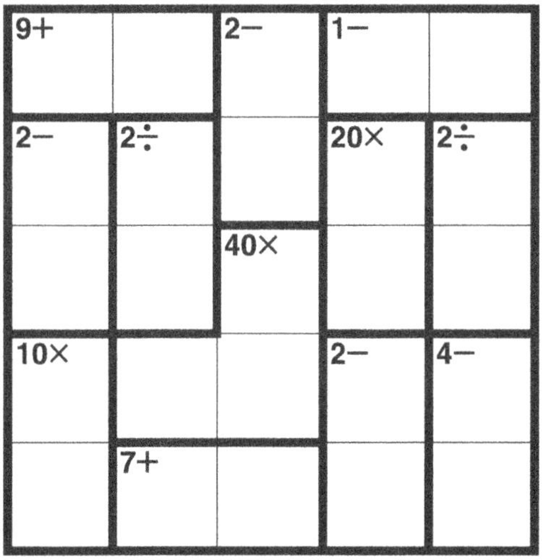

10 + − × ÷

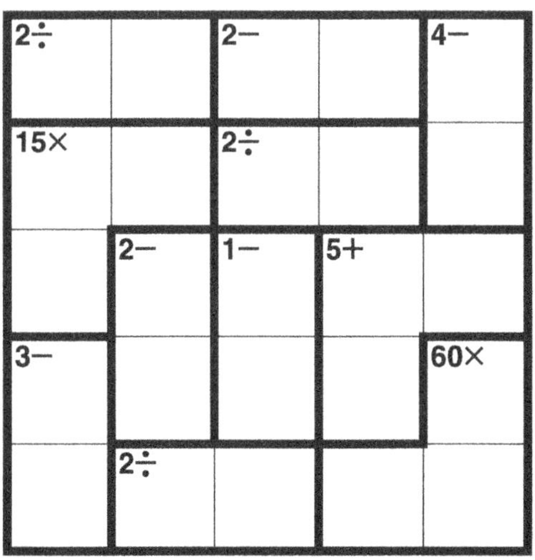

11 + − × ÷

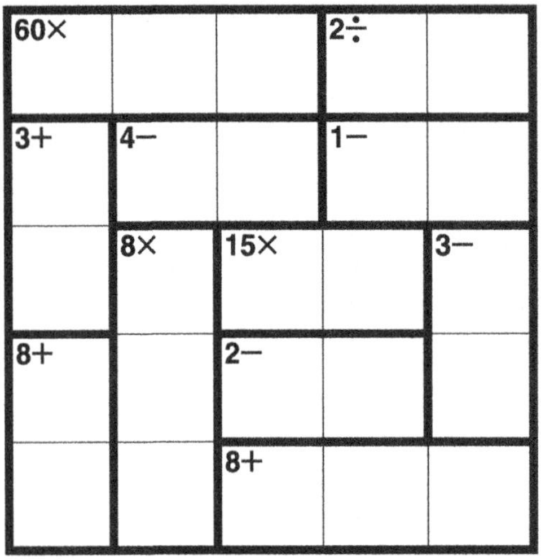

12 + − × ÷

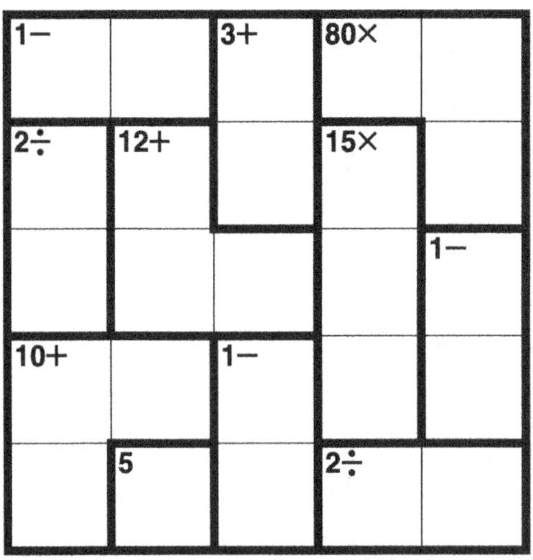

13 + − × ÷

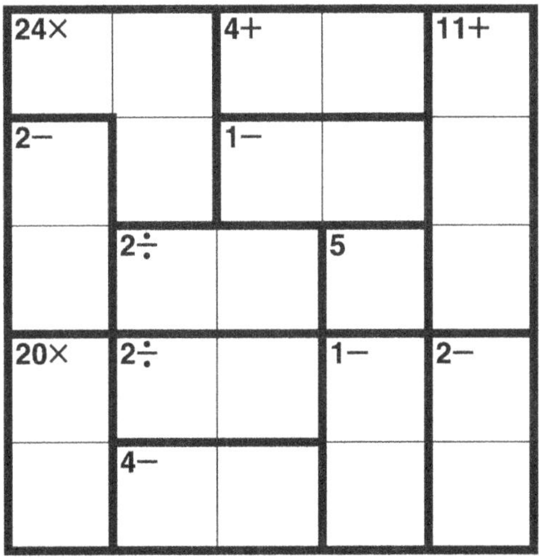

14 + − × ÷

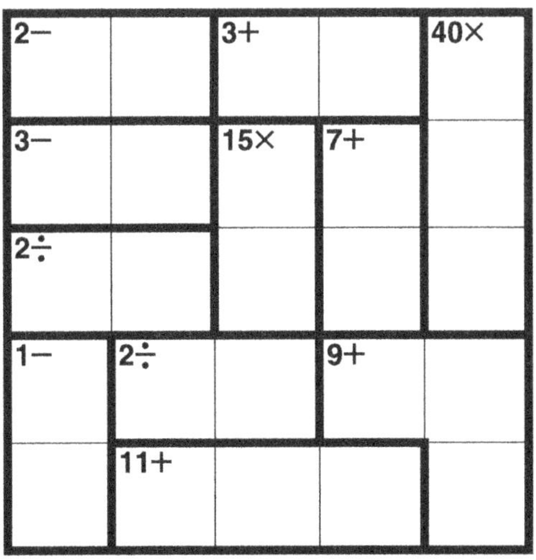

15 + − × ÷

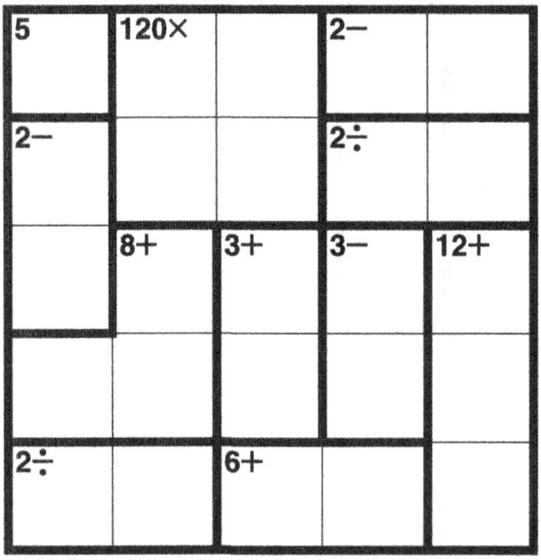

16 + − × ÷

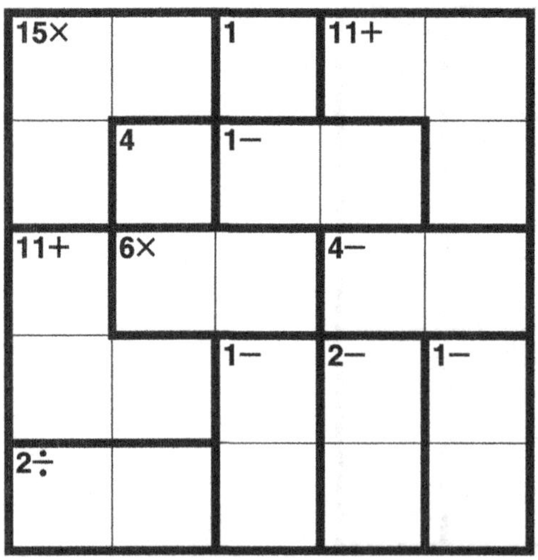

17 + − × ÷

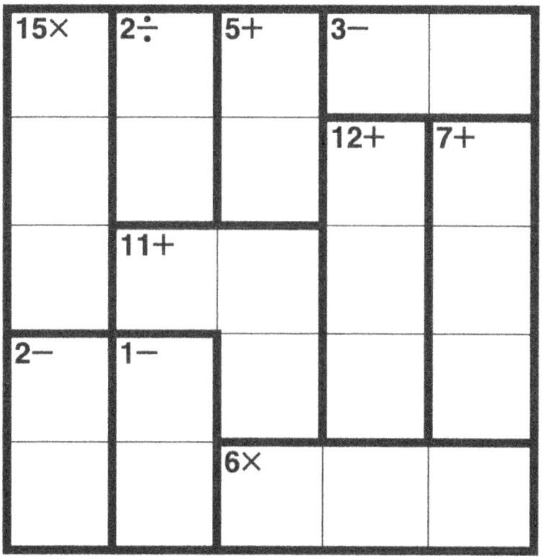

18 + − × ÷

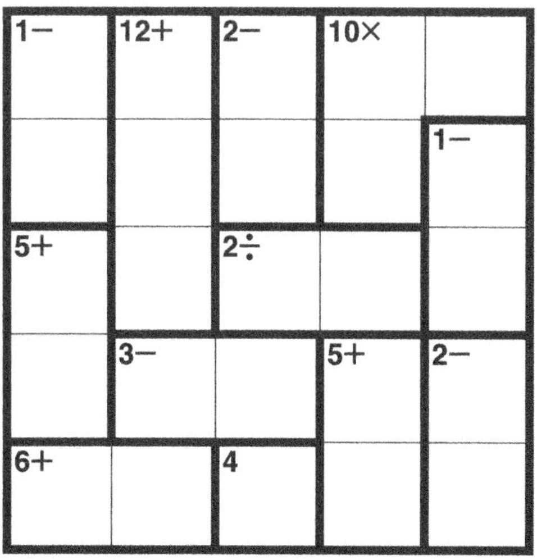

19 + − × ÷

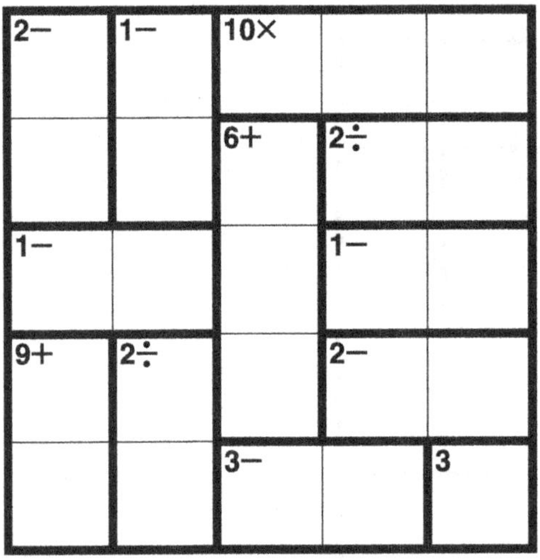

20 + − × ÷

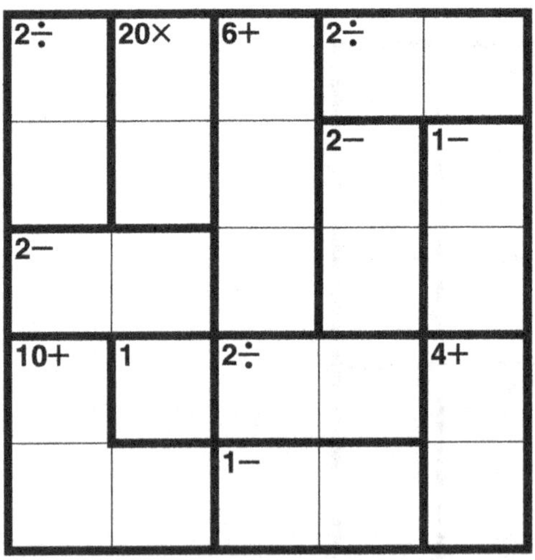

21 + − × ÷

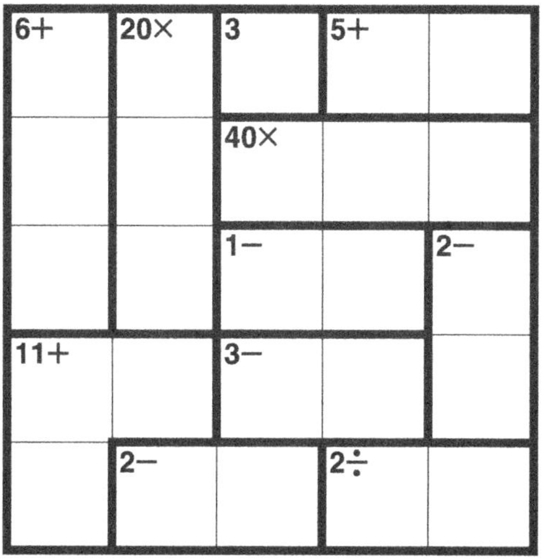

22 + − × ÷

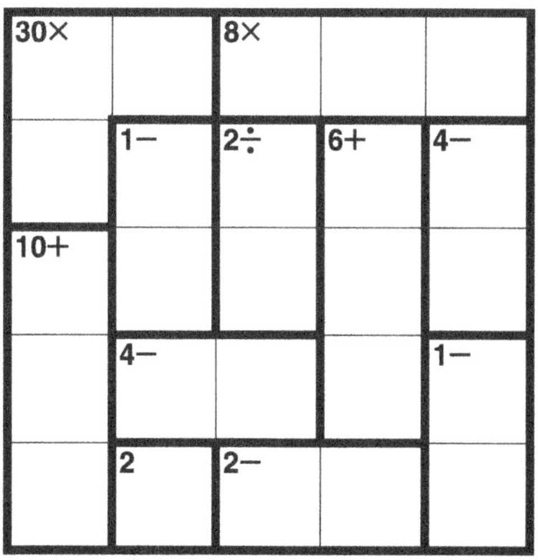

23 + − × ÷

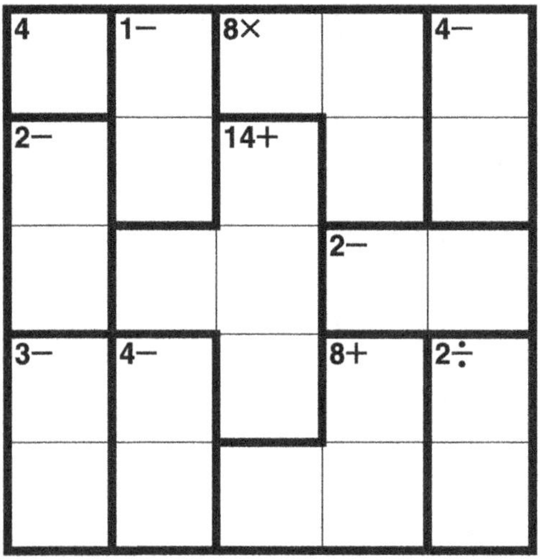

24 + − × ÷

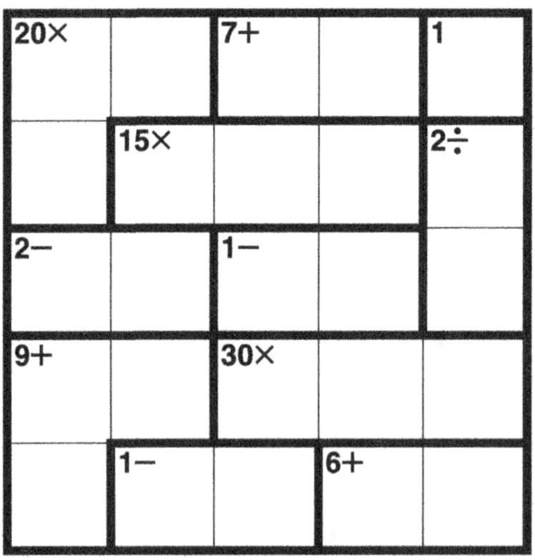

25 + − × ÷

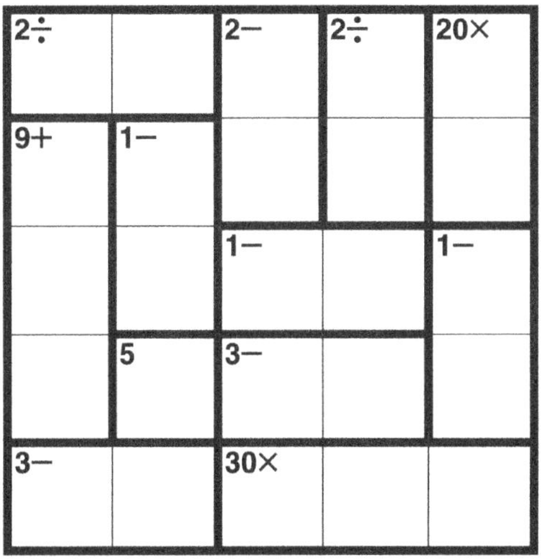

26 + − × ÷

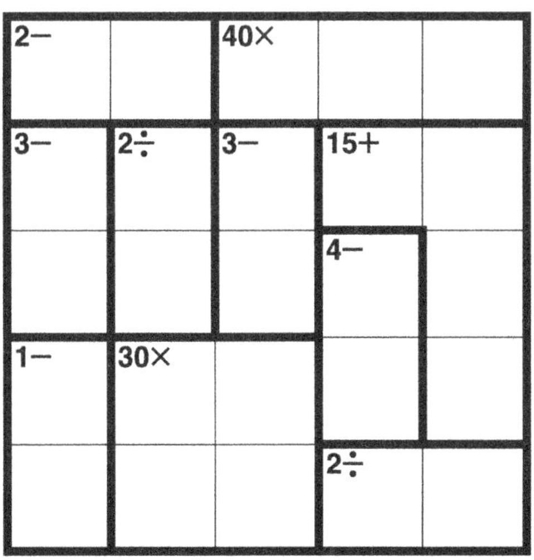

27 + − × ÷

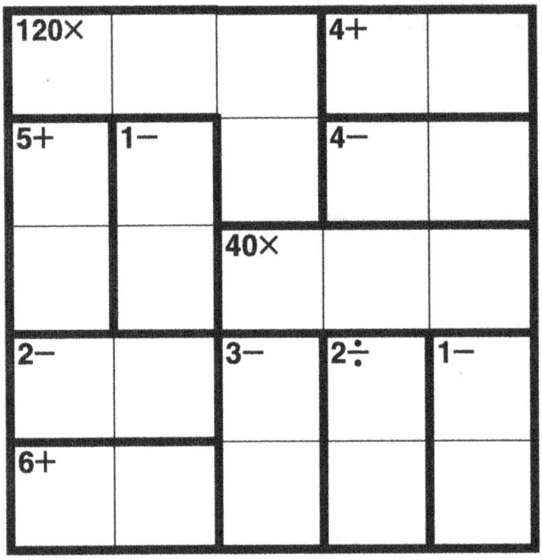

28 + − × ÷

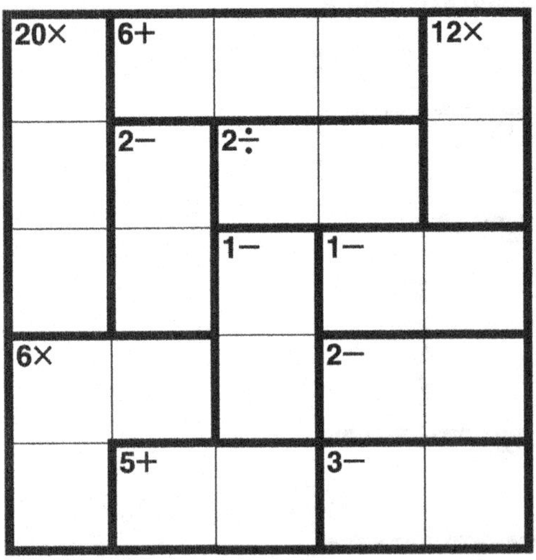

29 + − × ÷

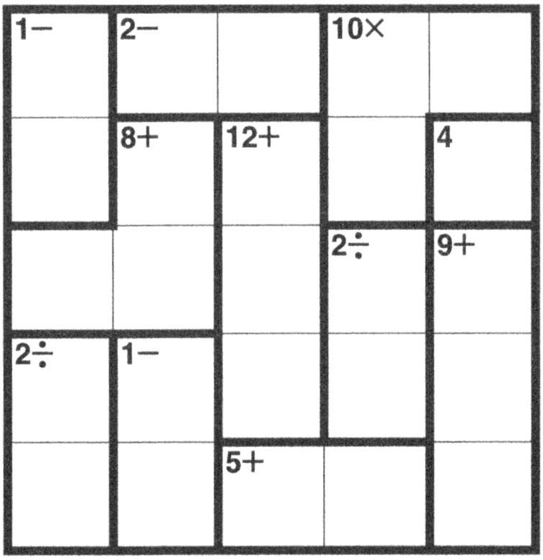

30 + − × ÷

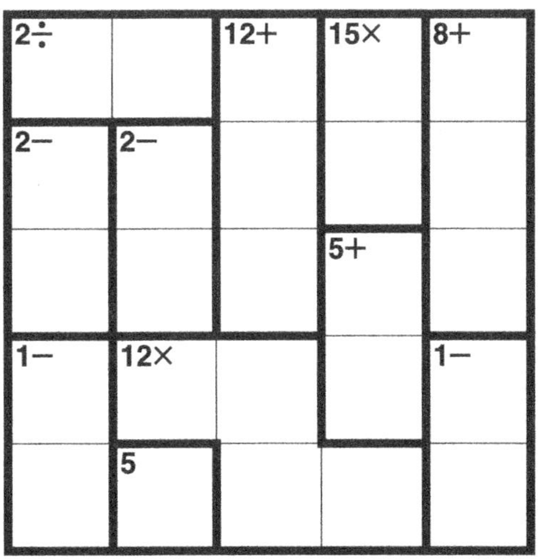

31 + − × ÷

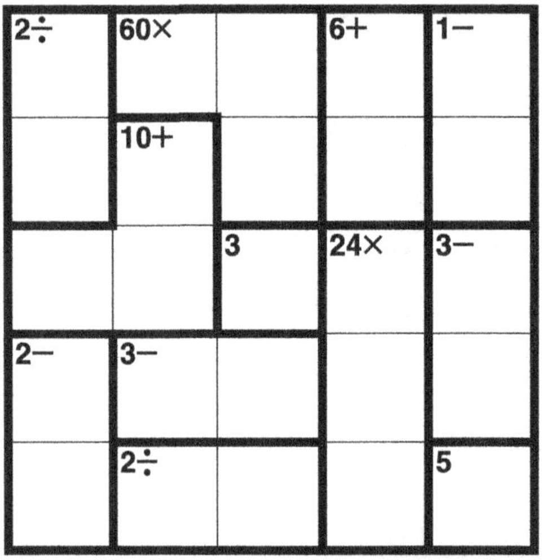

32 + − × ÷

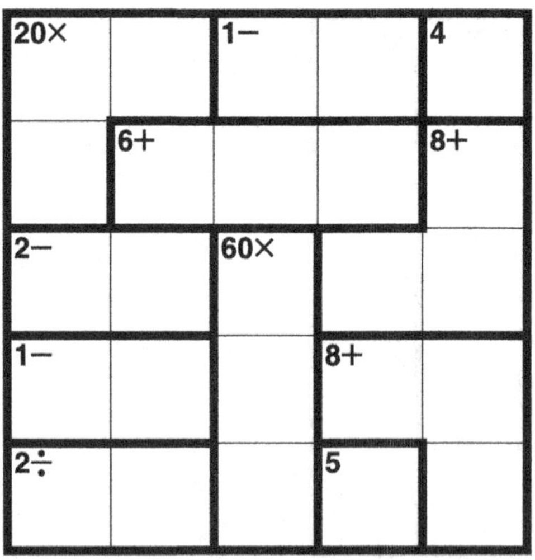

33 + − × ÷

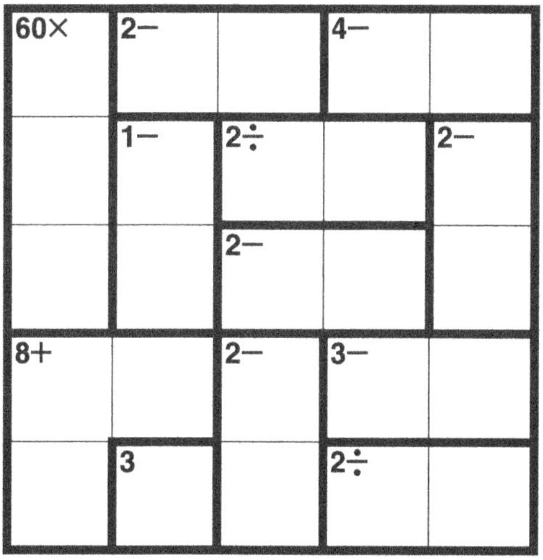

34 + − × ÷

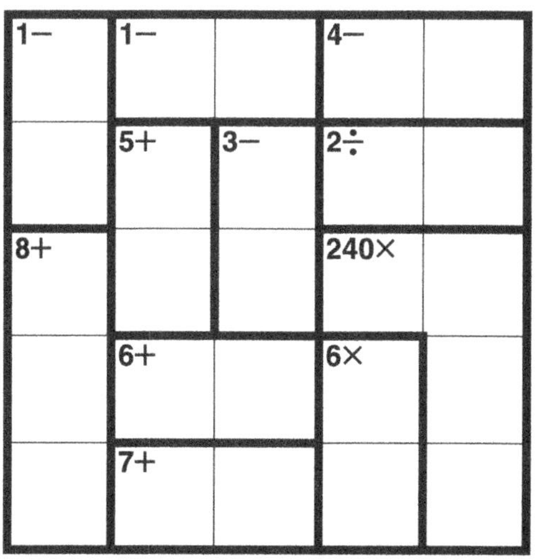

35 + − × ÷

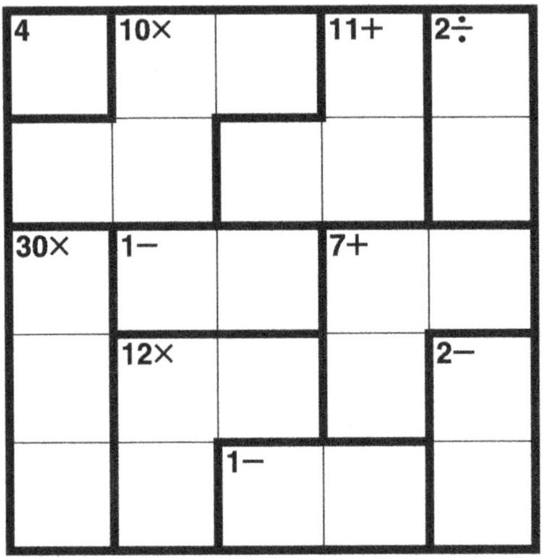

36 + − × ÷

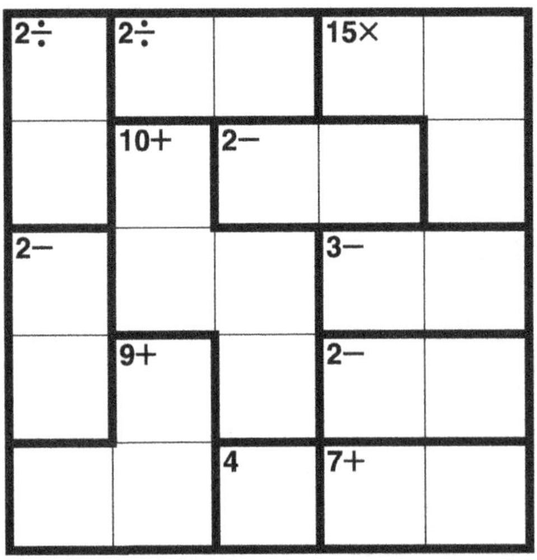

37 + − × ÷

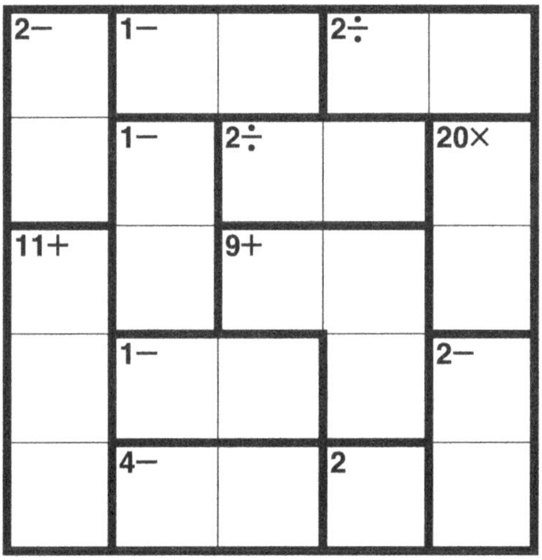

38 + − × ÷

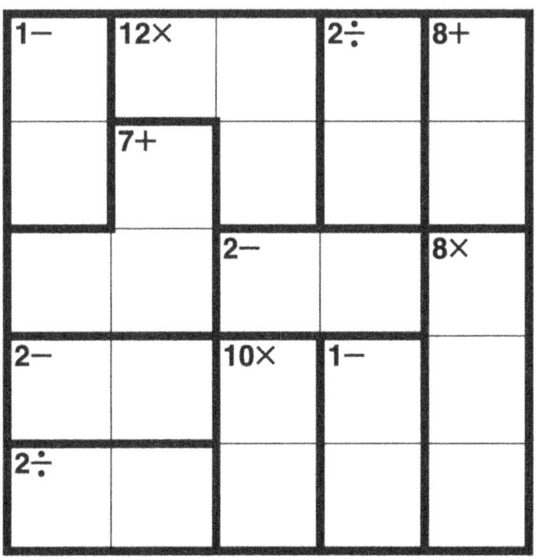

39 + − × ÷

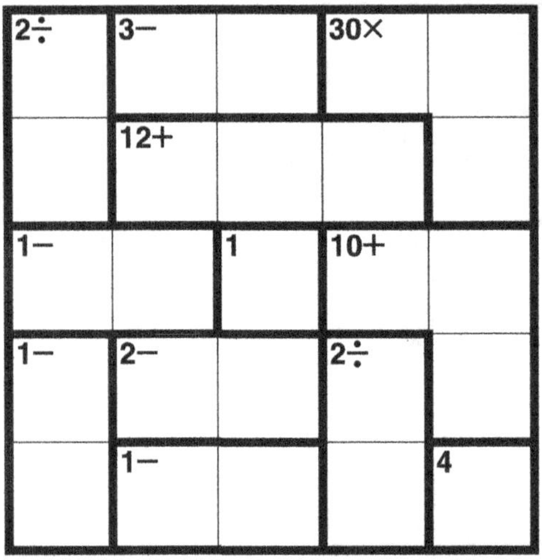

40 + − × ÷

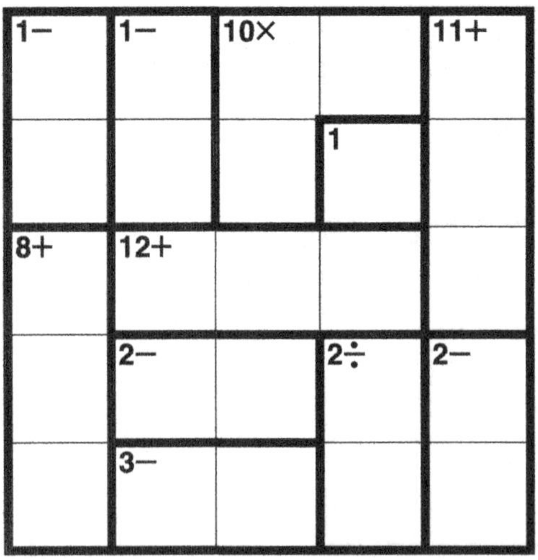

41 + − × ÷

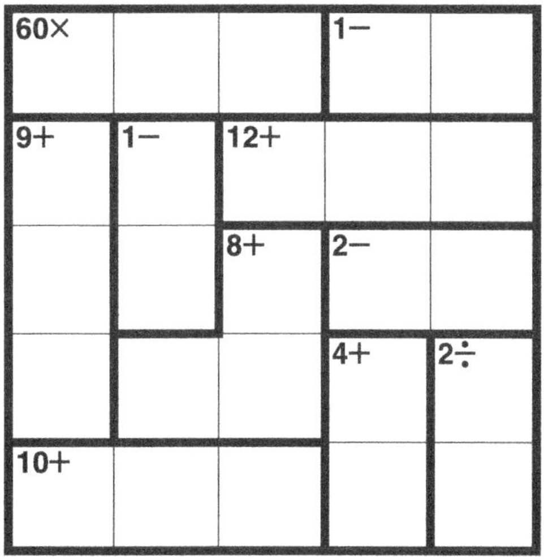

42 + − × ÷

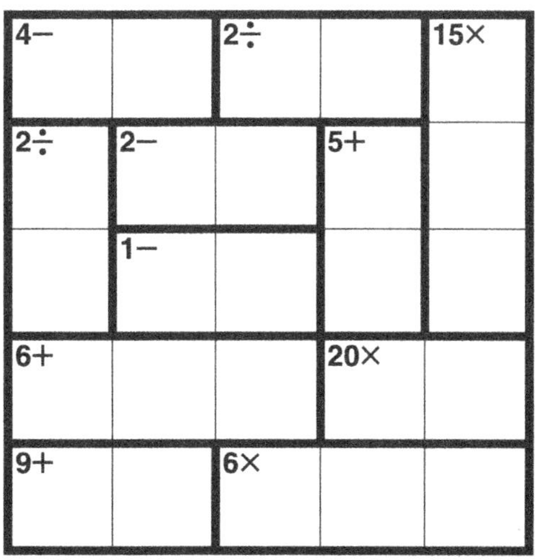

43 + − × ÷

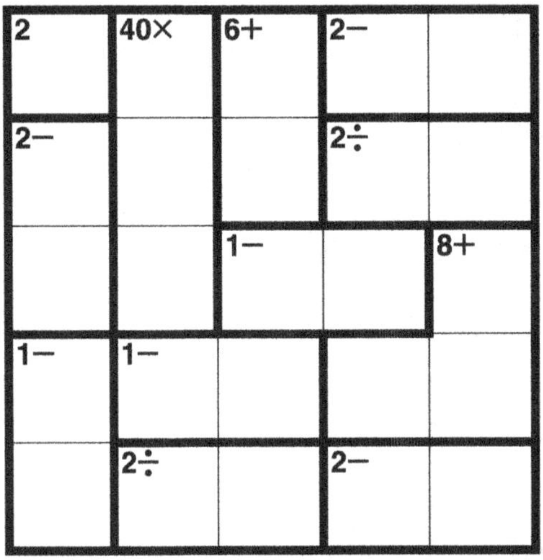

44 + − × ÷

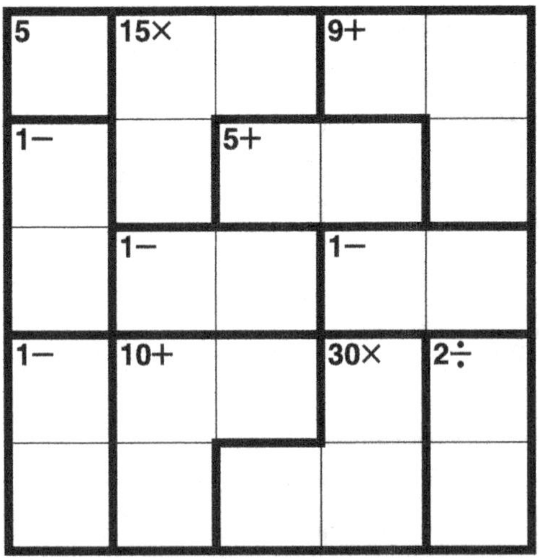

45

+ − × ÷

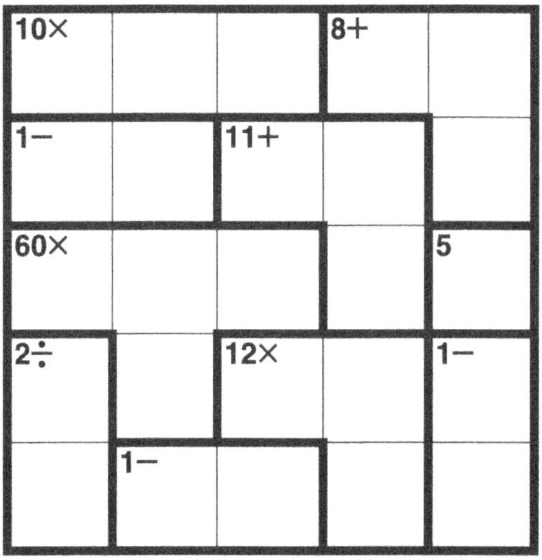

46

+ − × ÷

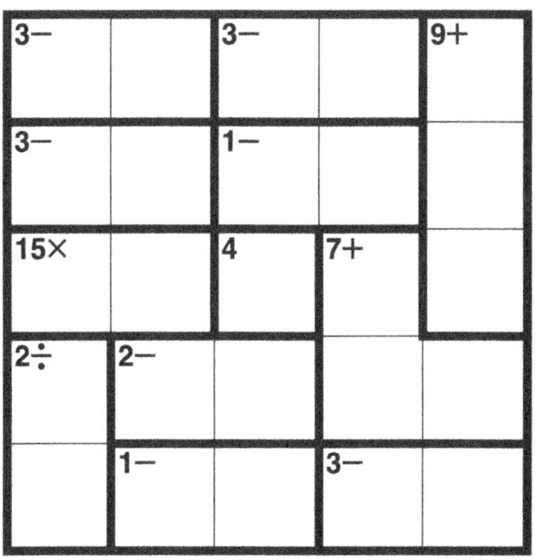

47 + − × ÷

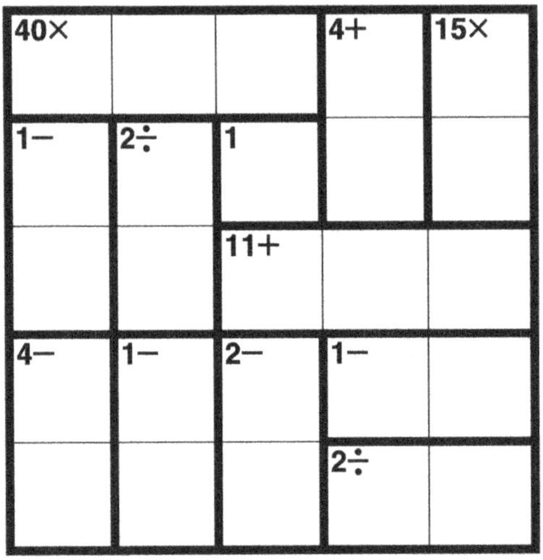

48 + − × ÷

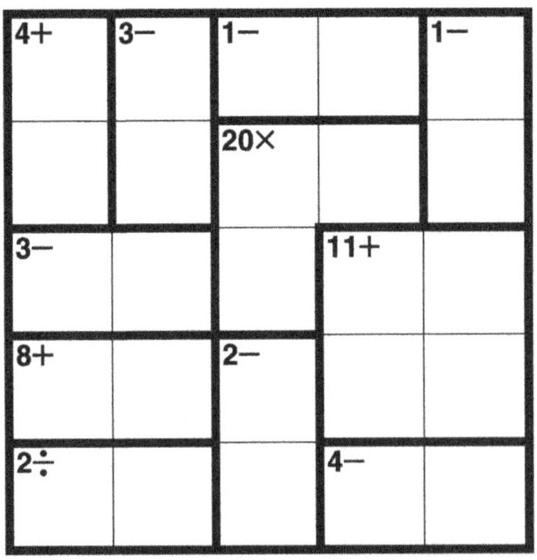

49

+ − × ÷

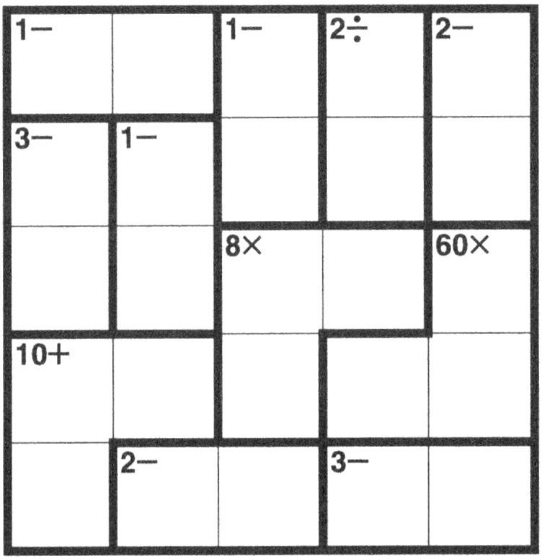

50

+ − × ÷

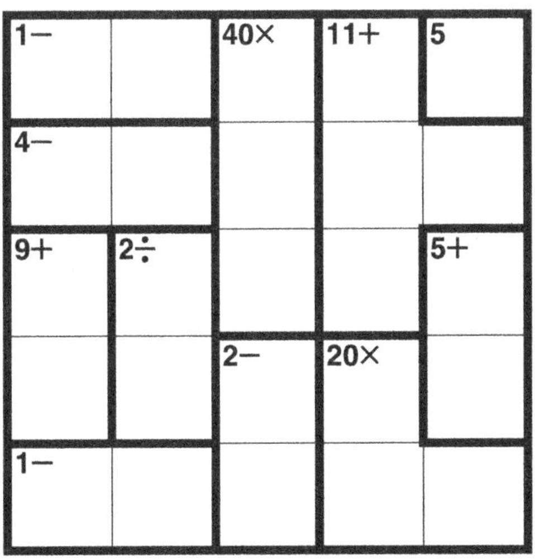

51 + − × ÷

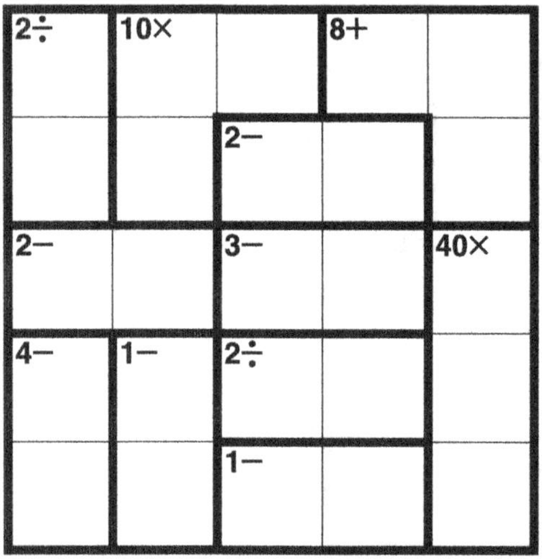

52 + − × ÷

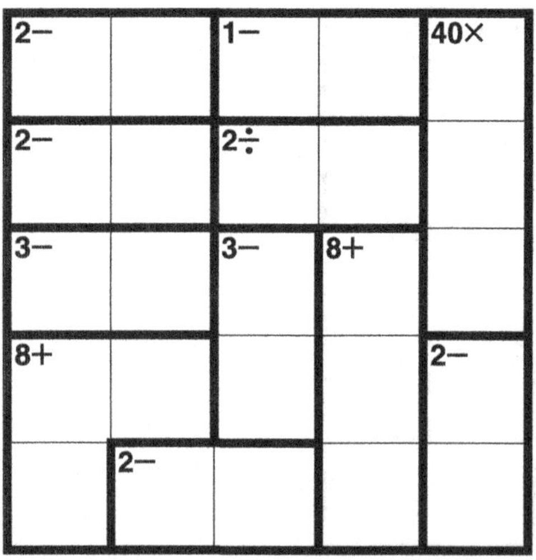

53 + − × ÷

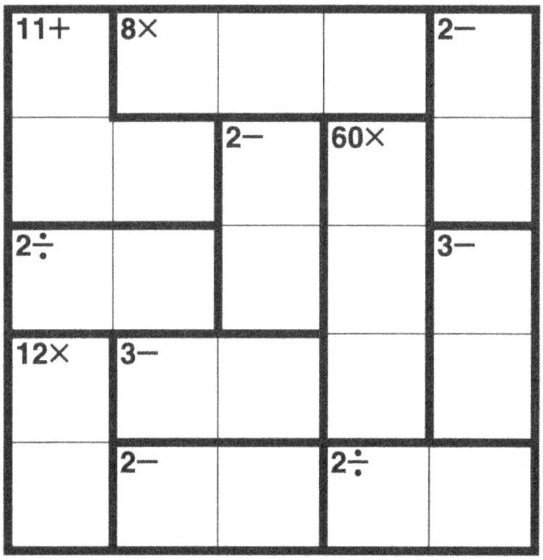

54 + − × ÷

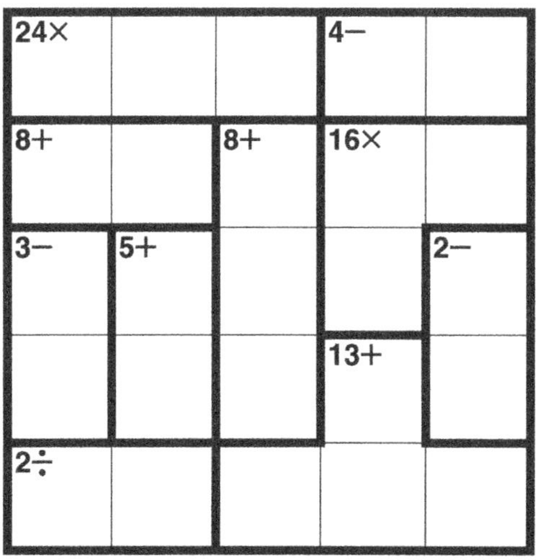

55

+ − × ÷

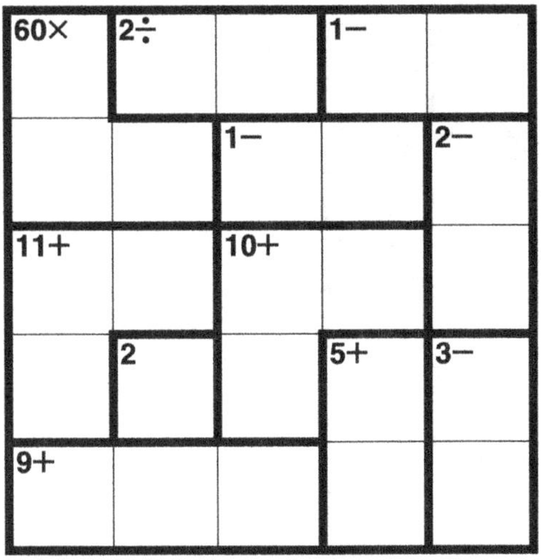

56

+ − × ÷

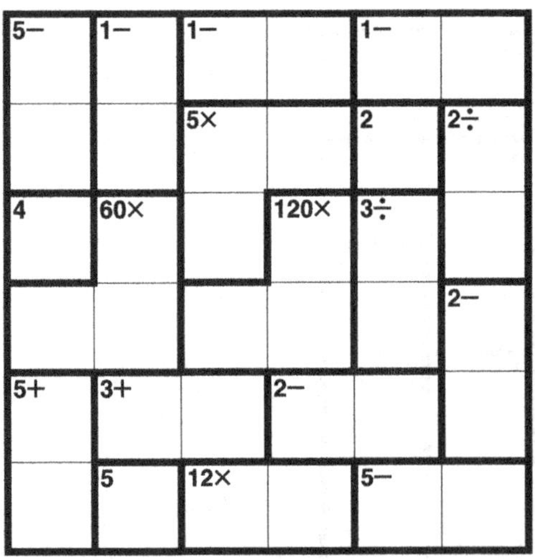

57 + − × ÷

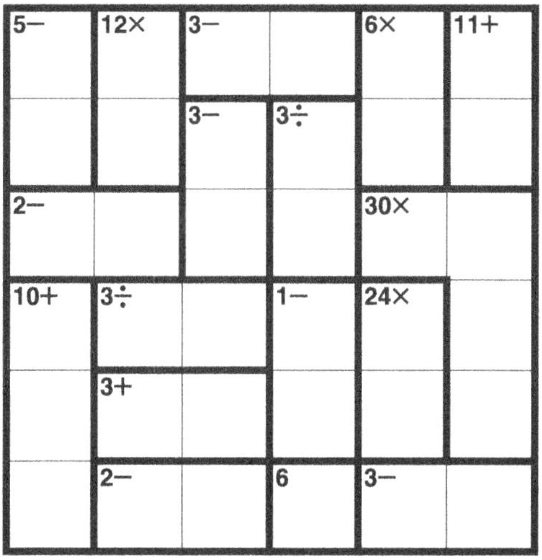

58 + − × ÷

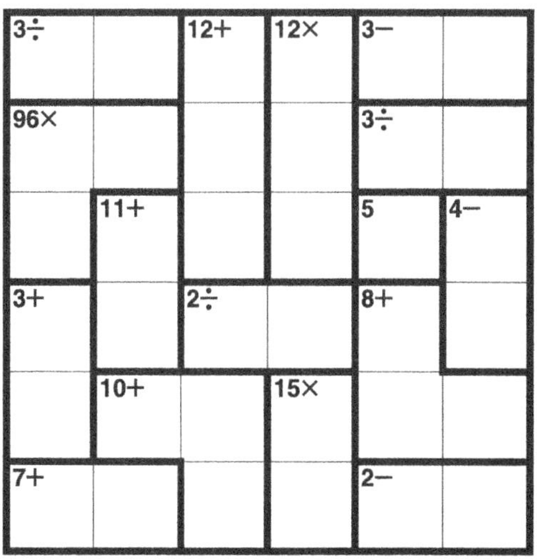

59 + − × ÷

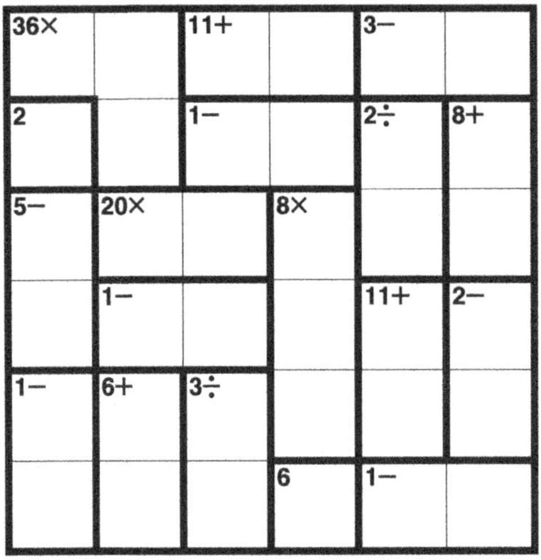

60 + − × ÷

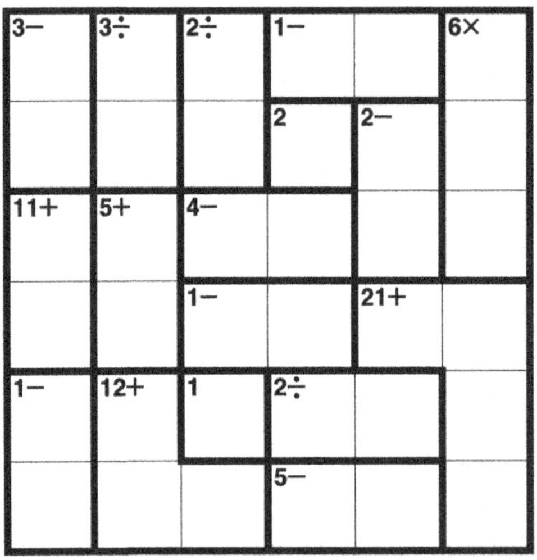

61 + − × ÷

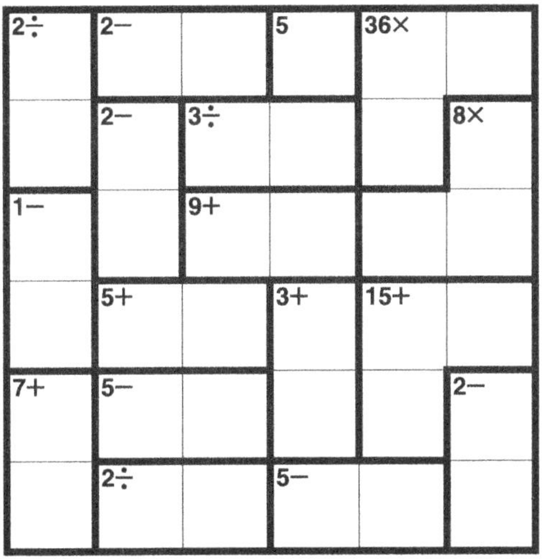

62 + − × ÷

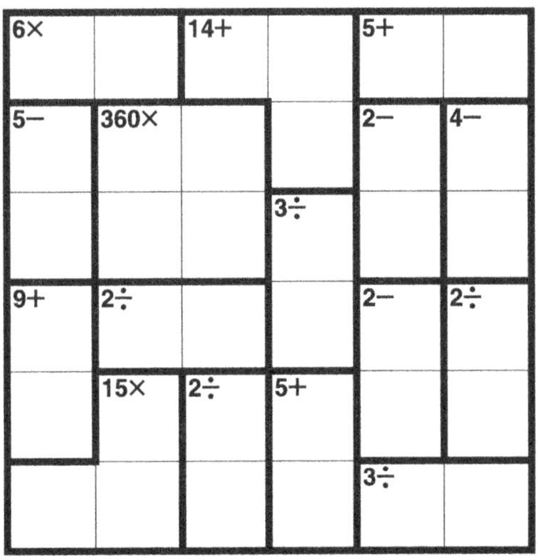

63

+ − × ÷

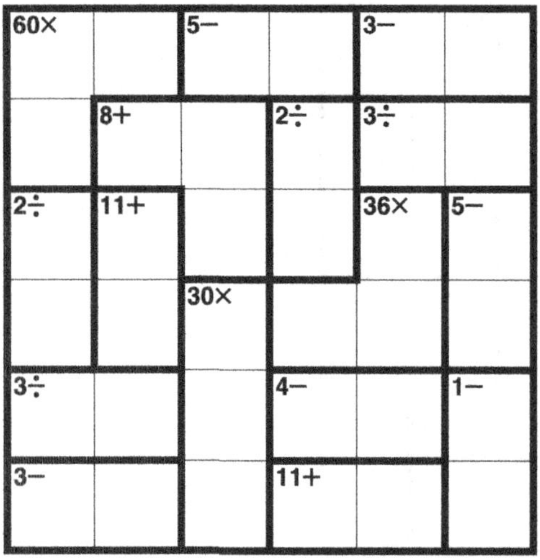

64

+ − × ÷

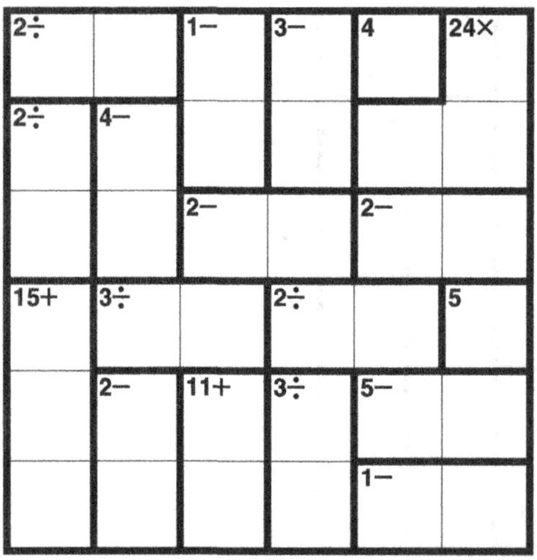

65 + − × ÷

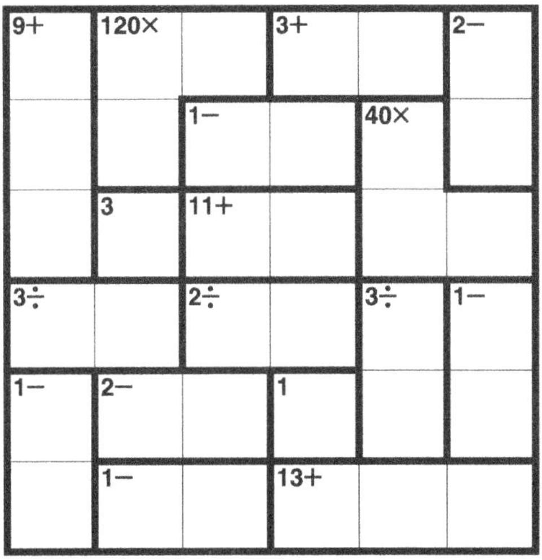

66 + − × ÷

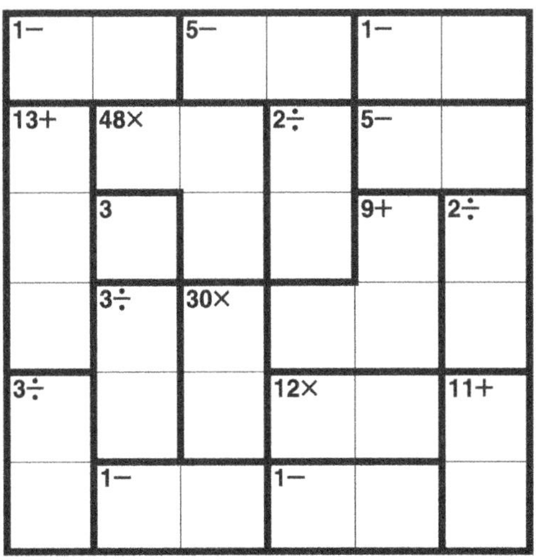

67 + − × ÷

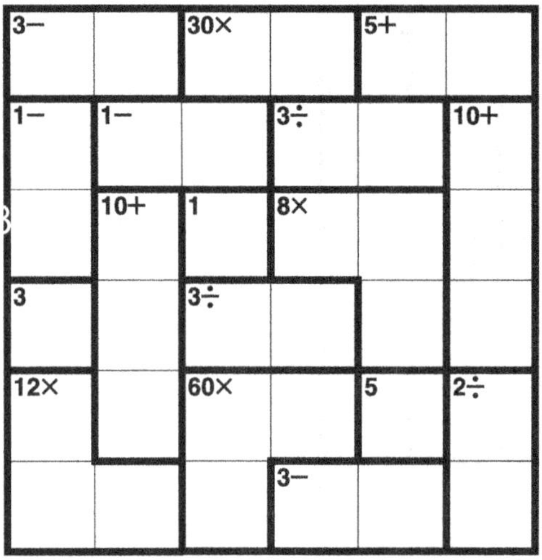

68 + − × ÷

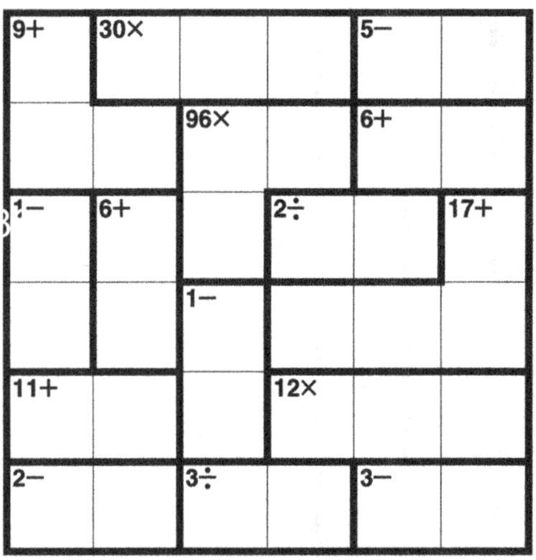

69

+ − × ÷

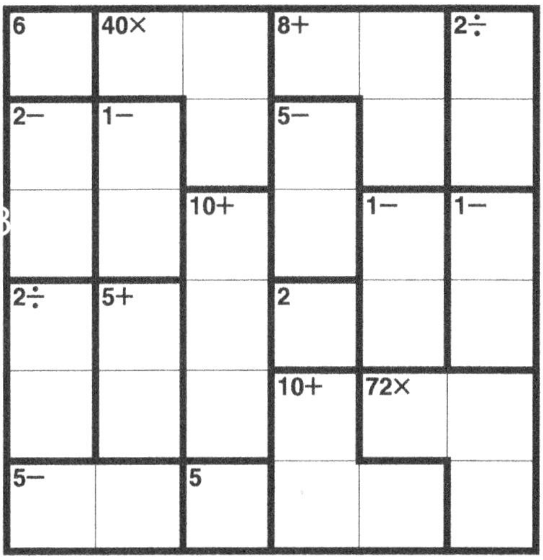

70

+ − × ÷

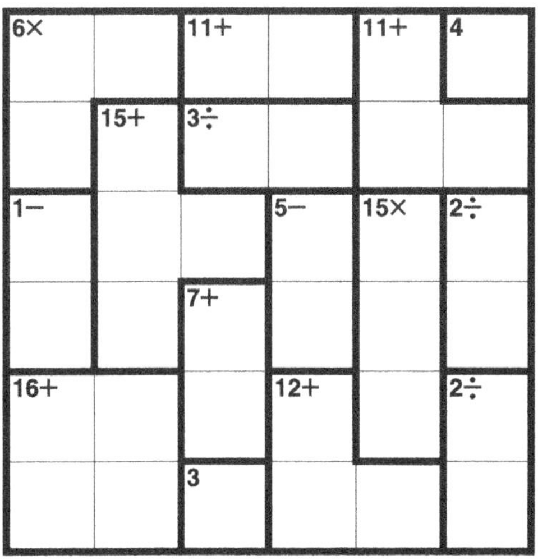

71 + − × ÷

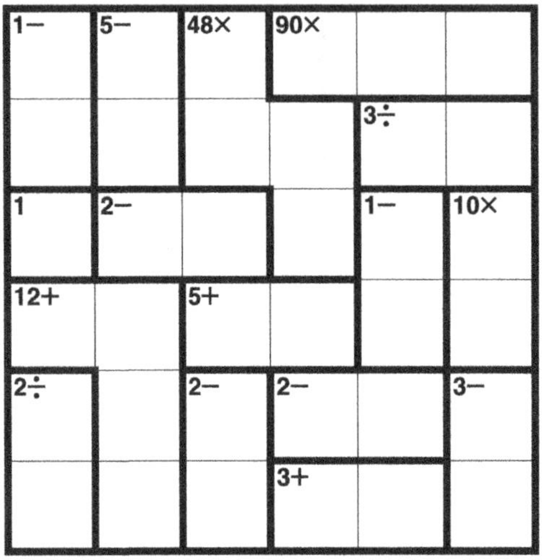

72 + − × ÷

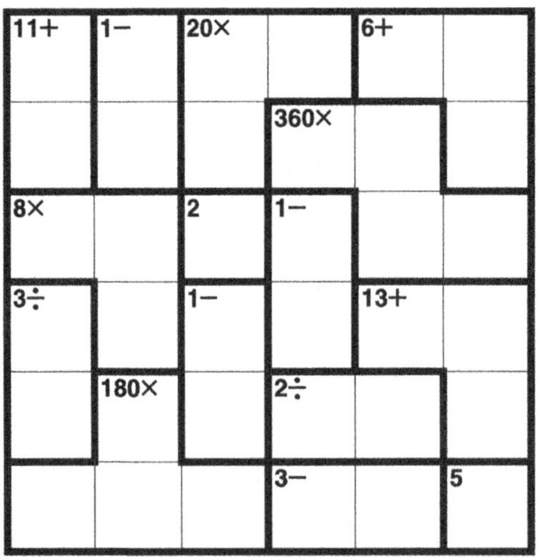

73

+ − × ÷

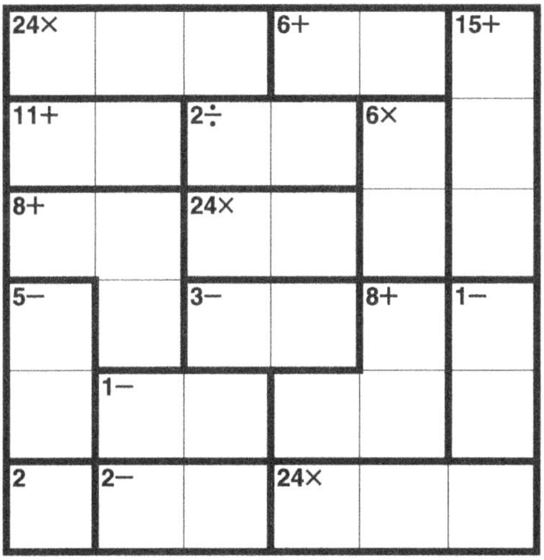

74

+ − × ÷

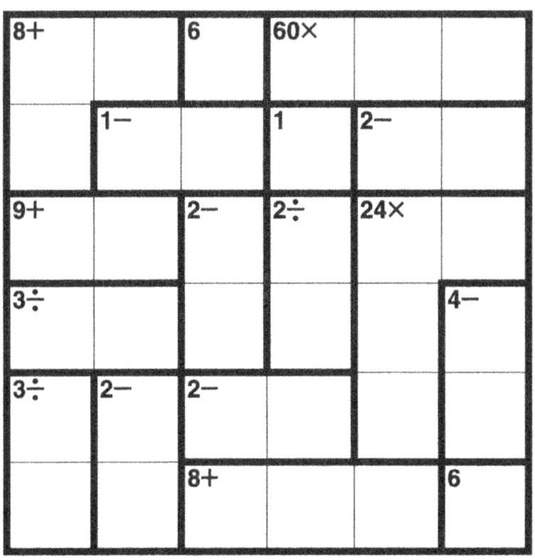

75 + − × ÷

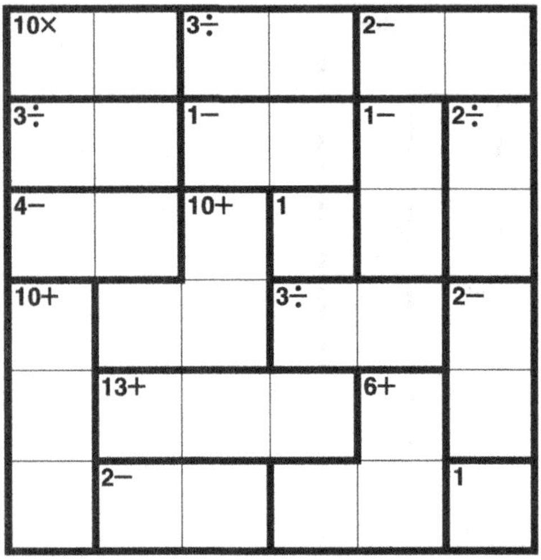

76 + − × ÷

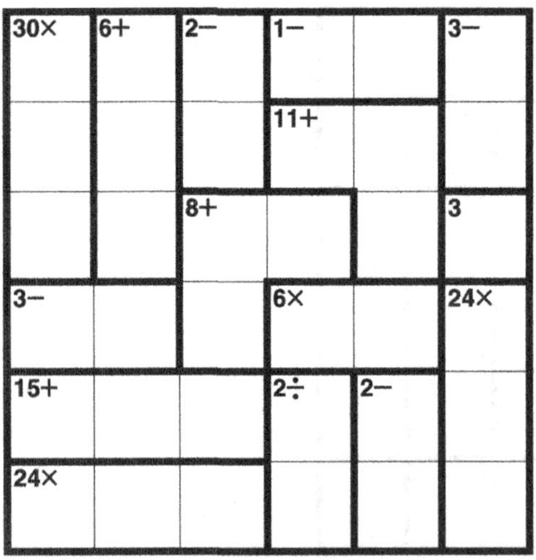

77 + − × ÷

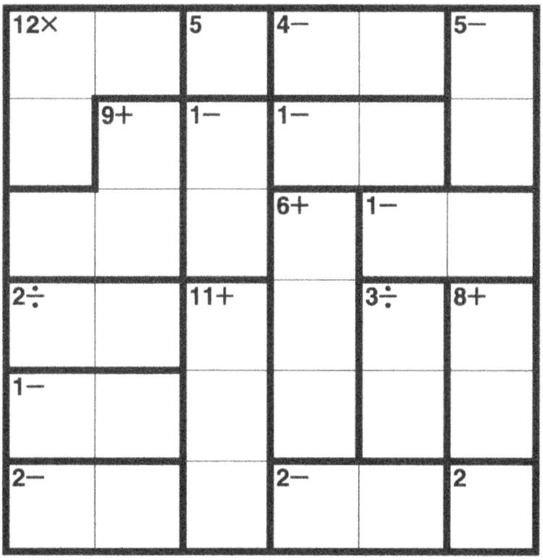

78 + − × ÷

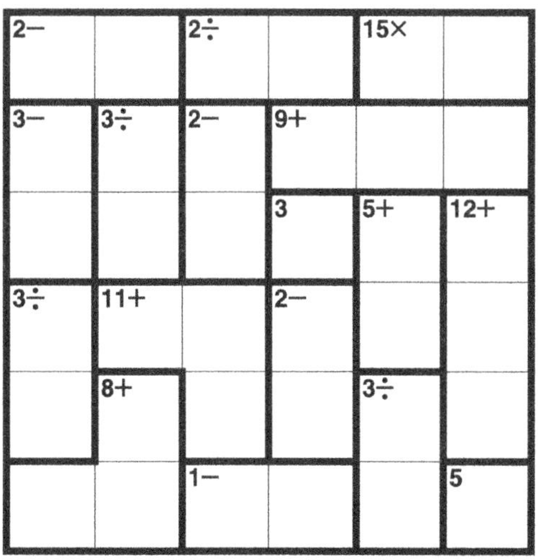

79 + − × ÷

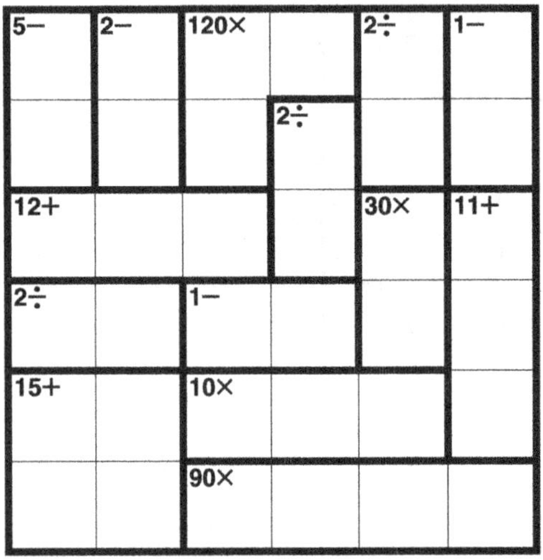

80 + − × ÷

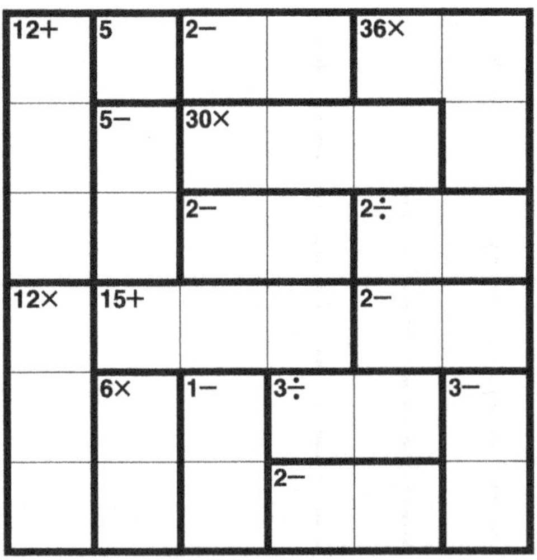

81 + − × ÷

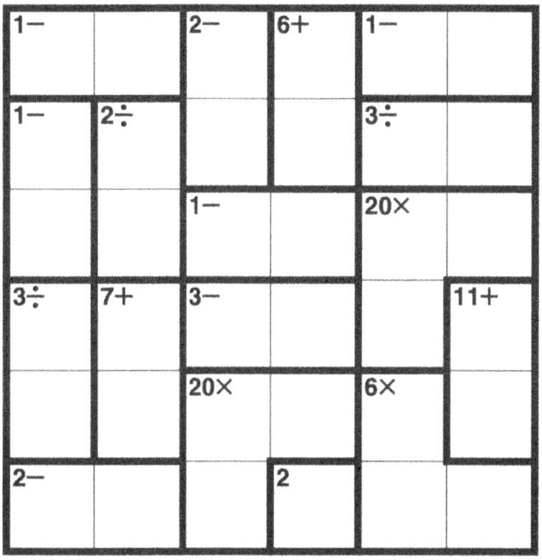

82 + − × ÷

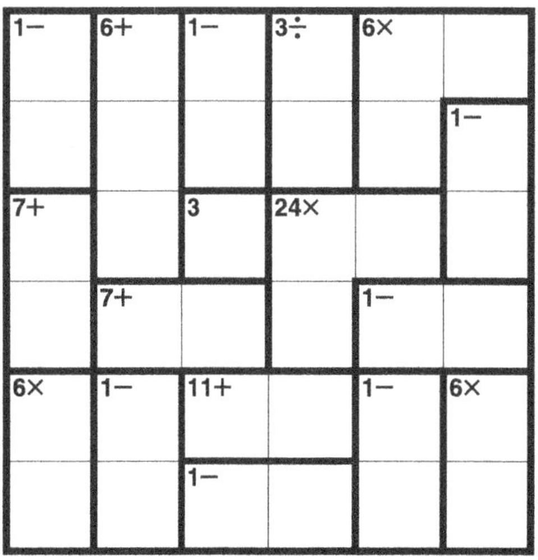

83 + − × ÷

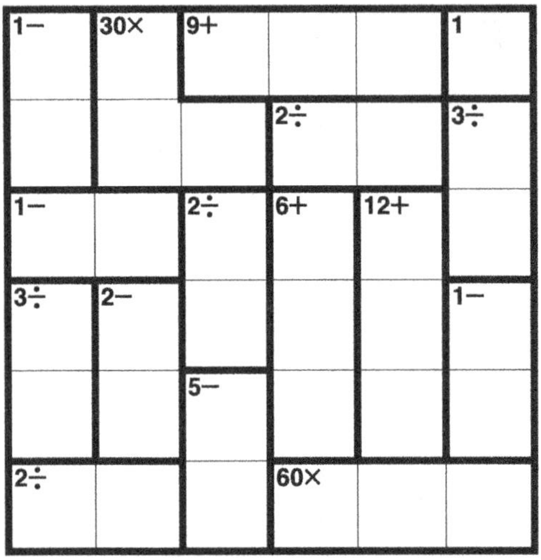

84 + − × ÷

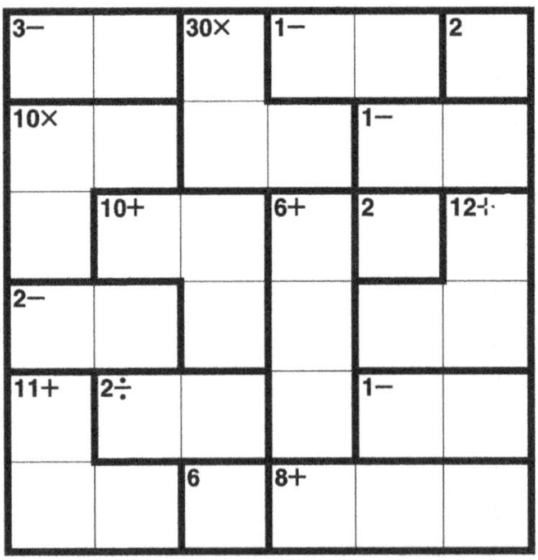

85

+ − × ÷

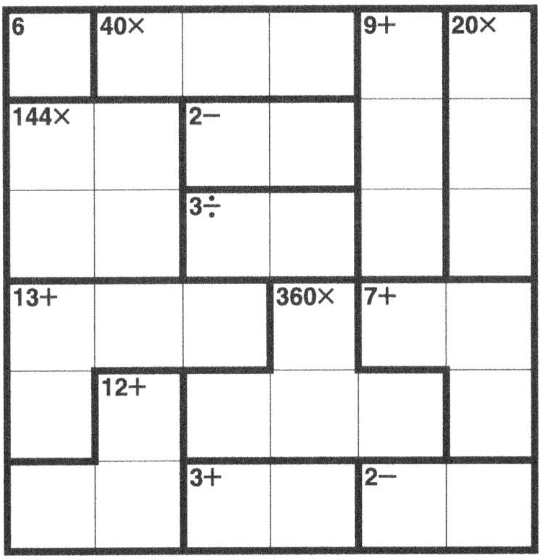

86

+ − × ÷

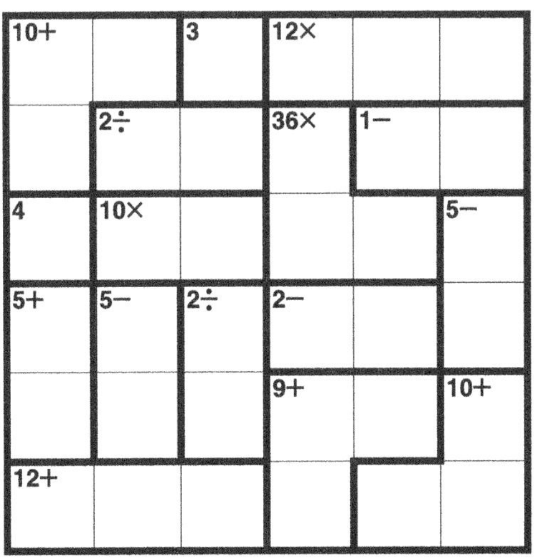

87 + − × ÷

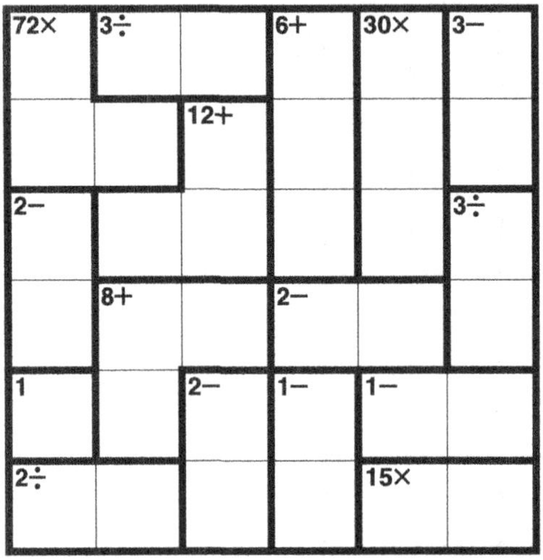

88 + − × ÷

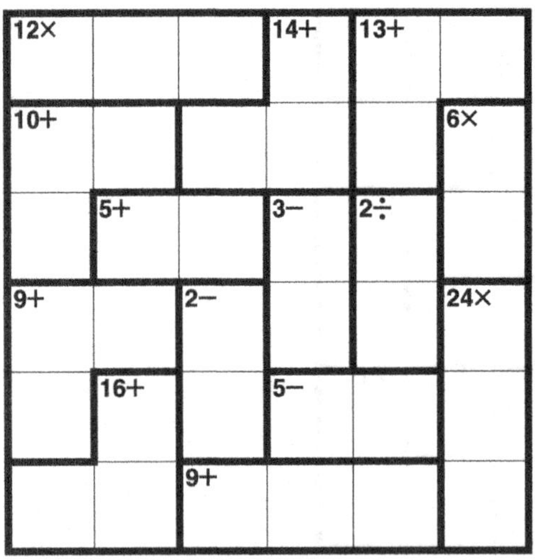

89 + − × ÷

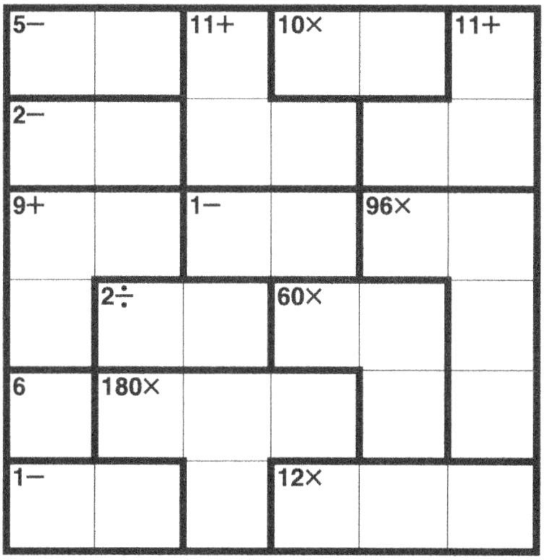

90 + − × ÷

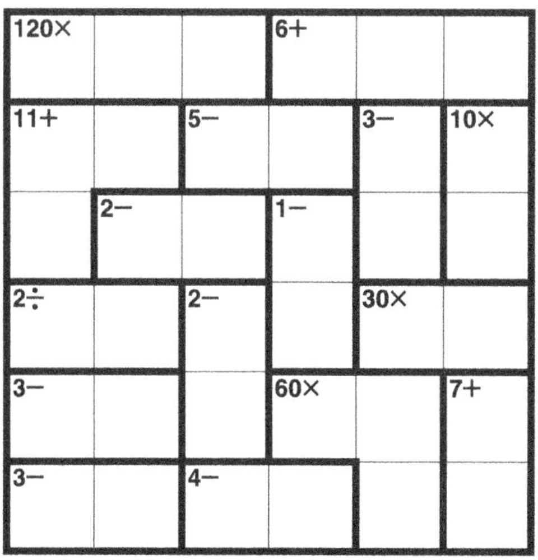

91 + − × ÷

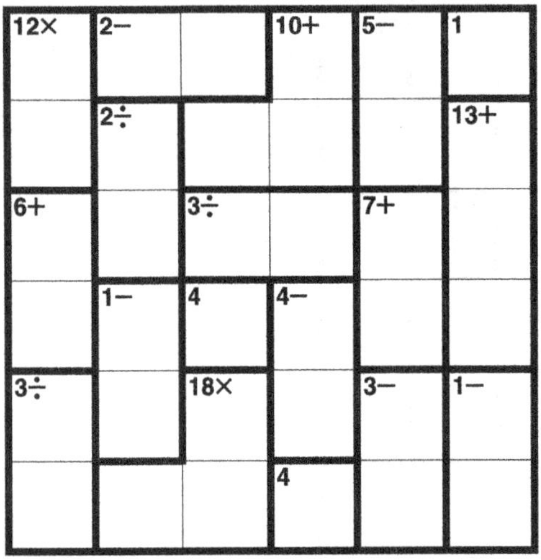

92 + − × ÷

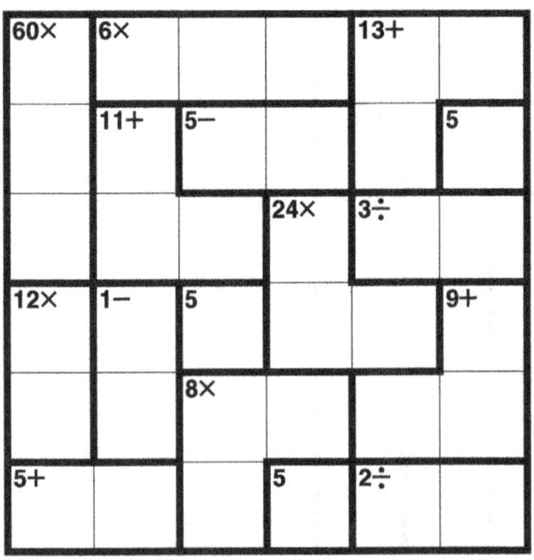

93 + − × ÷

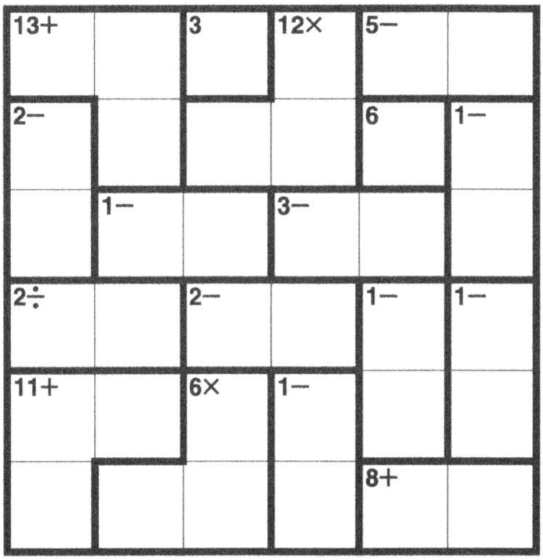

94 + − × ÷

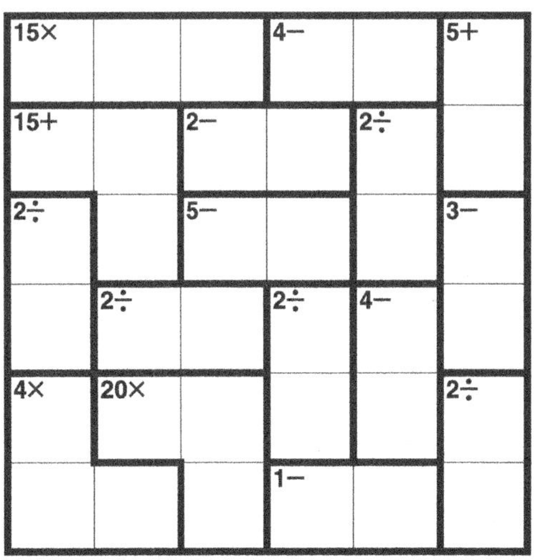

95 + − × ÷

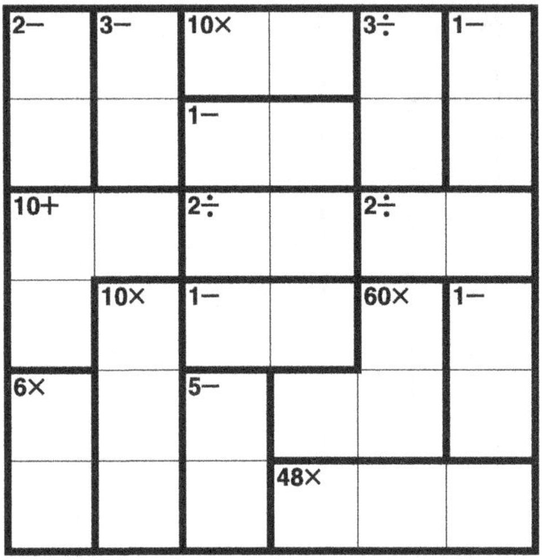

96 + − × ÷

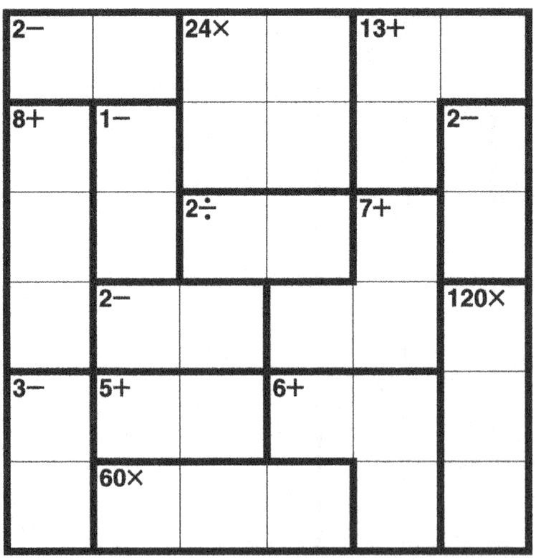

97 + − × ÷

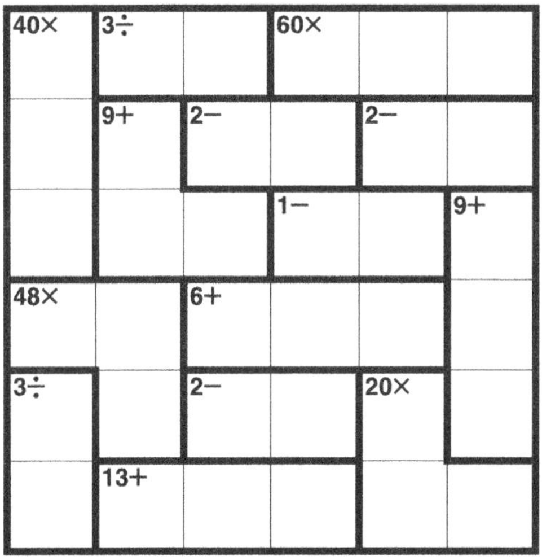

98 + − × ÷

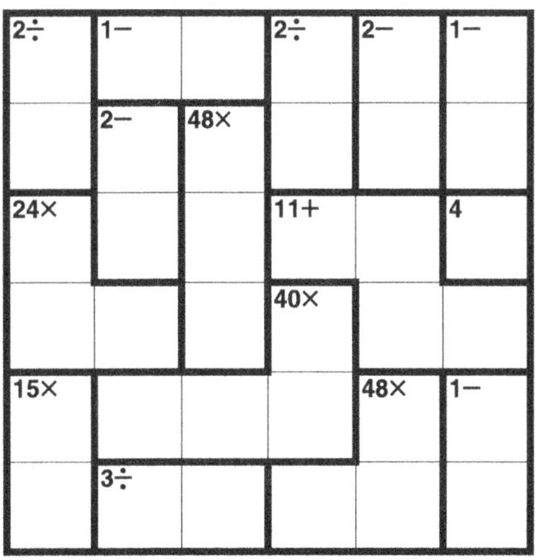

99 + − × ÷

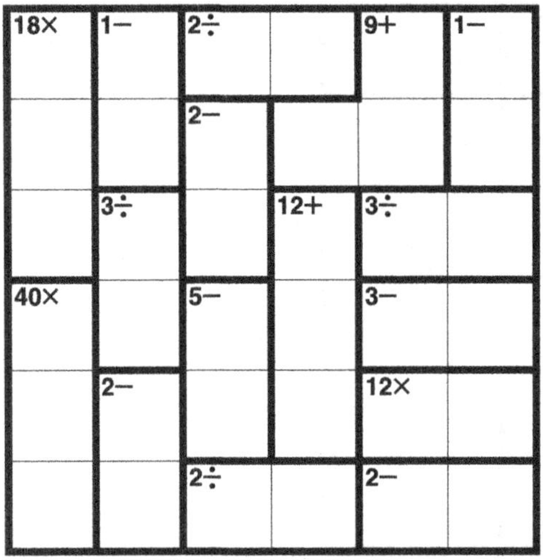

100 + − × ÷

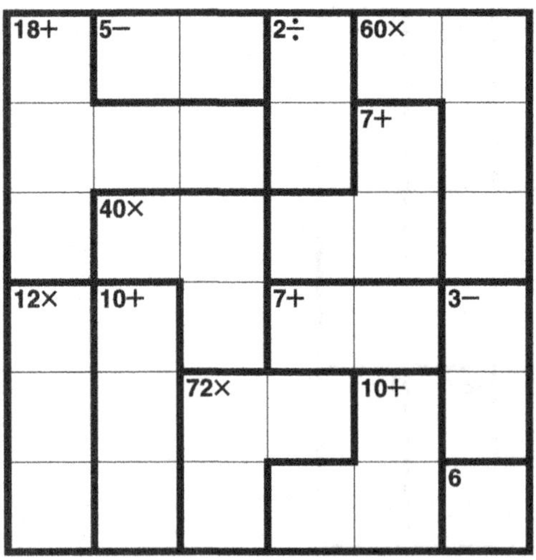

101 + − × ÷

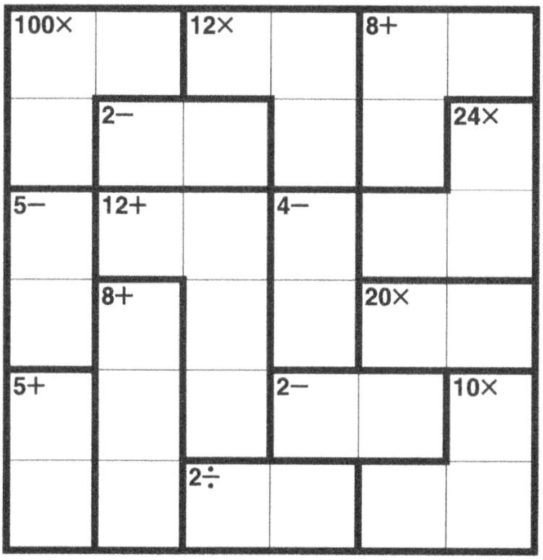

102 + − × ÷

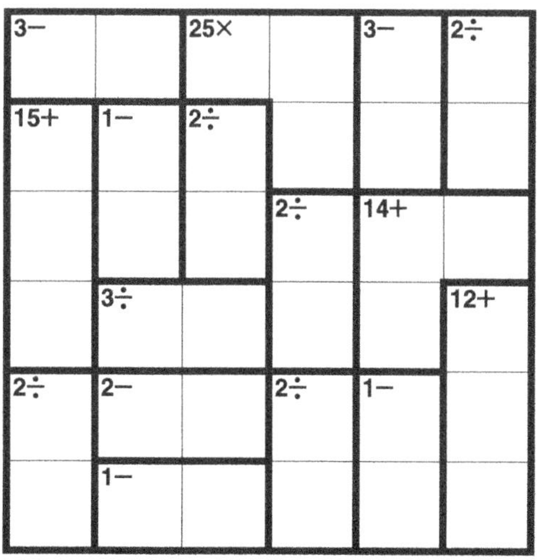

103 + − × ÷

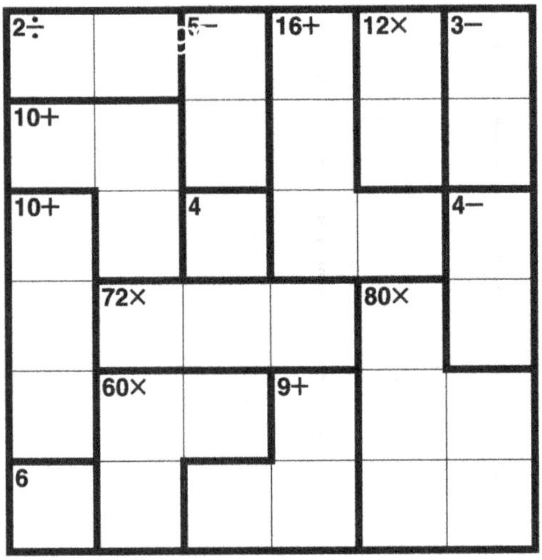

104 + − × ÷

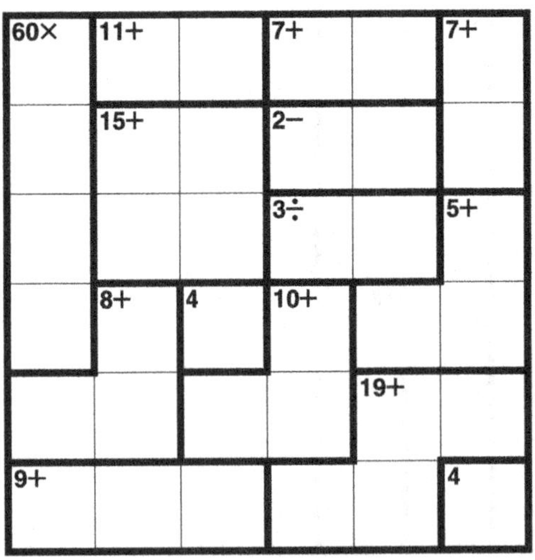

105 + − × ÷

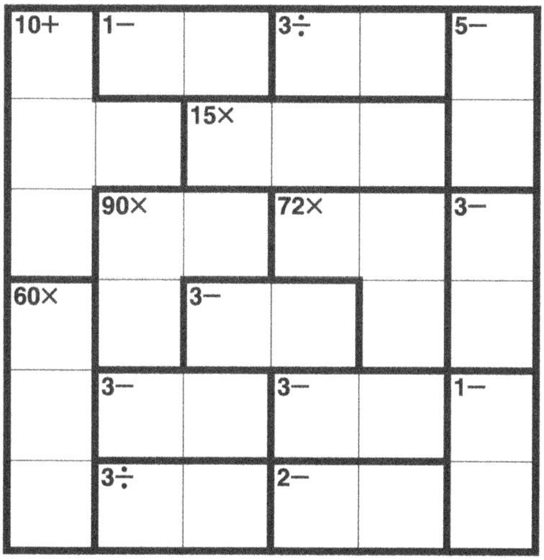

106 + − × ÷

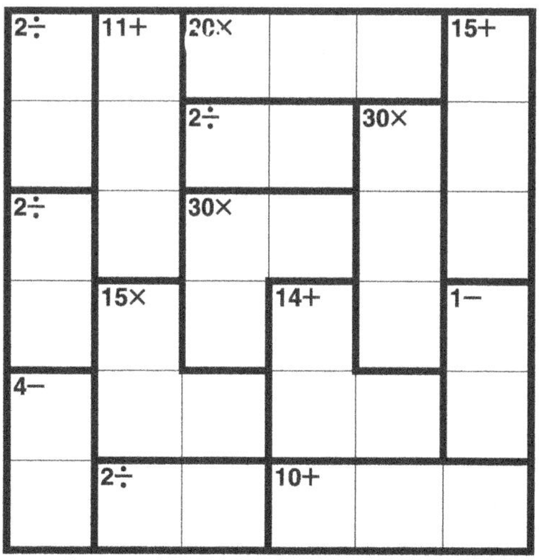

107 + − × ÷

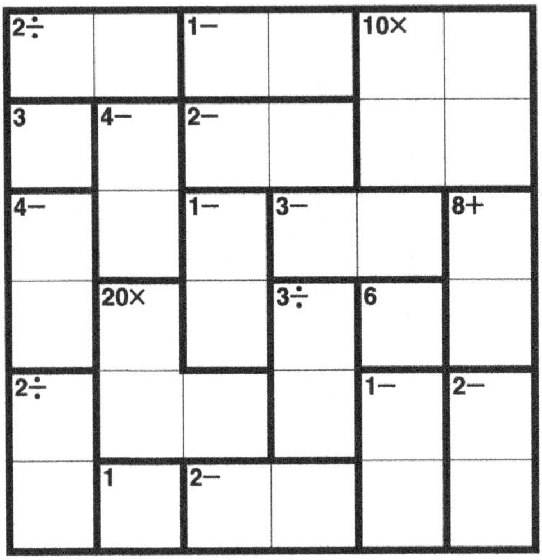

108 + − × ÷

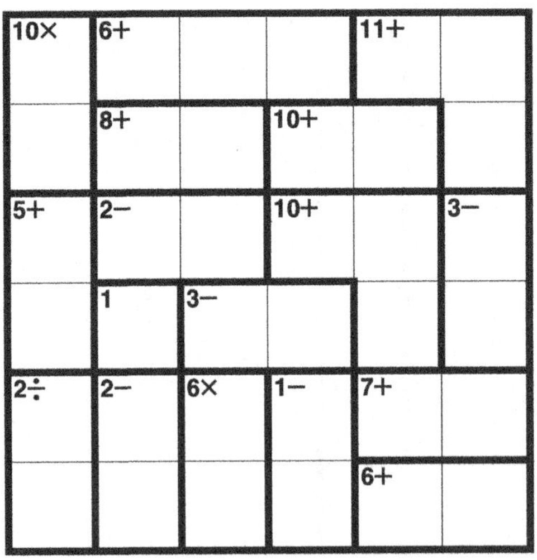

109 + − × ÷

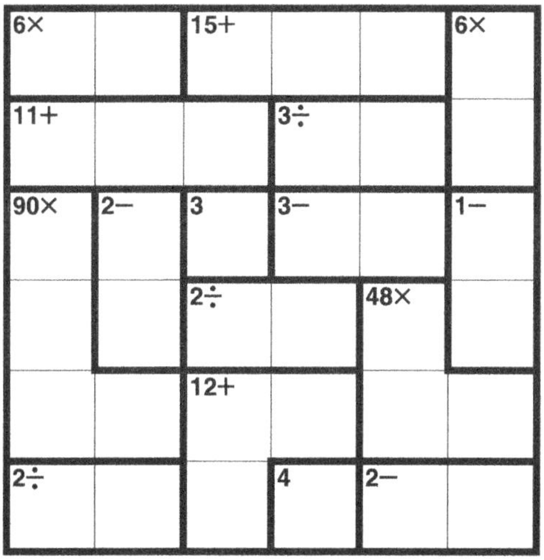

110 + − × ÷

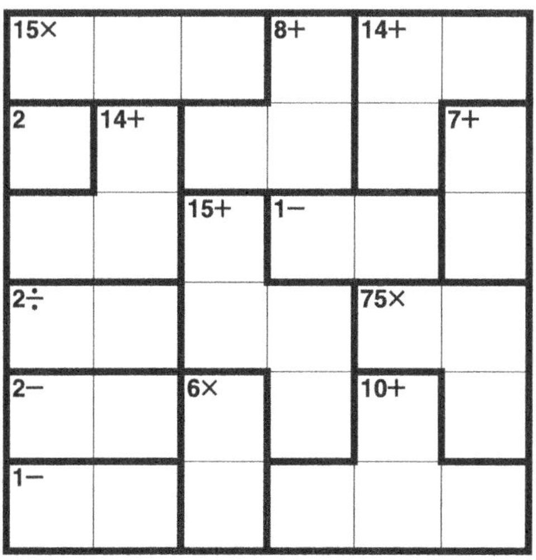

111 + − × ÷

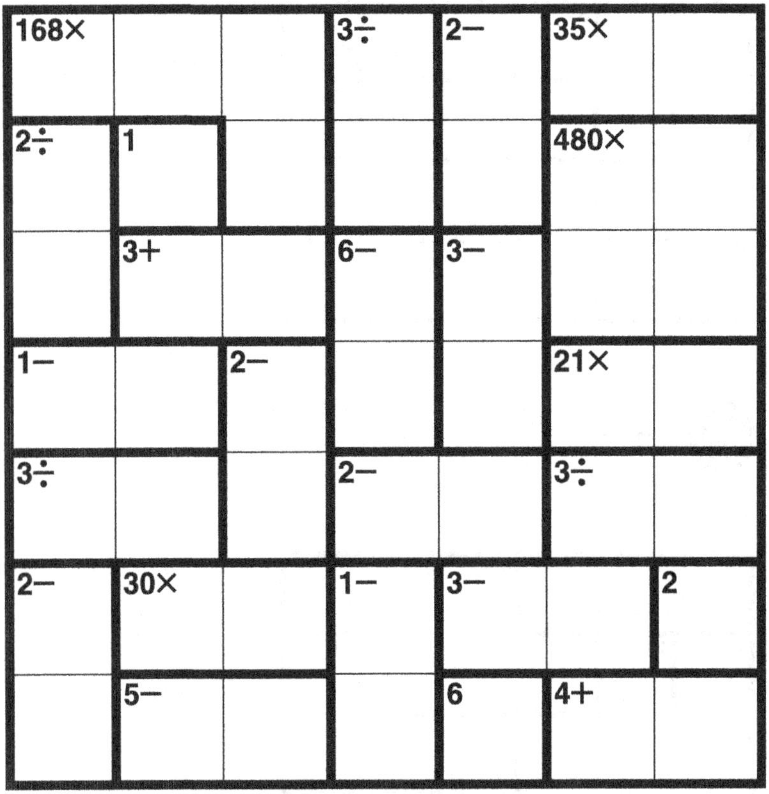

112 + − × ÷

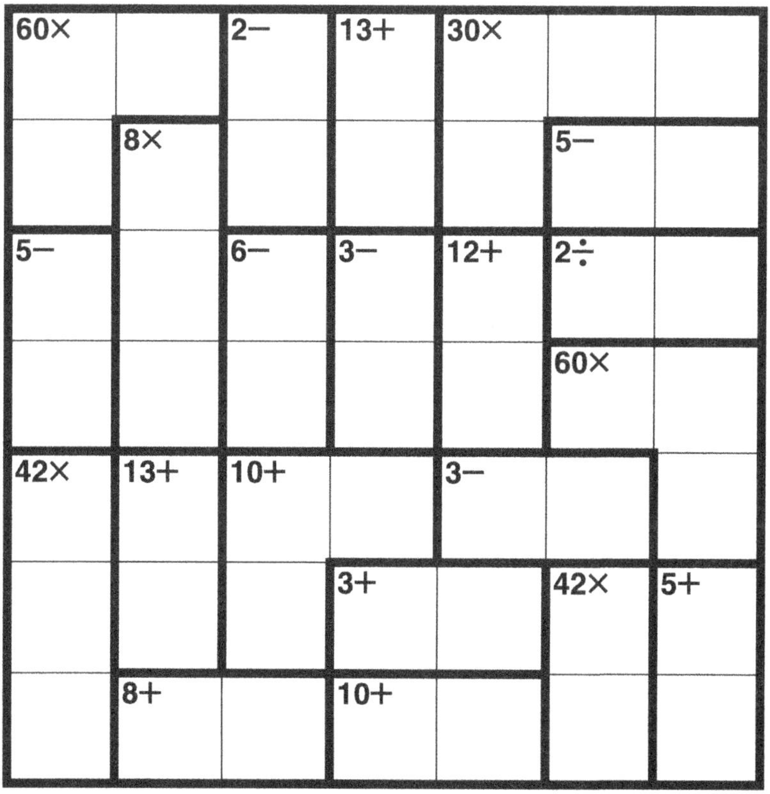

113 + − × ÷

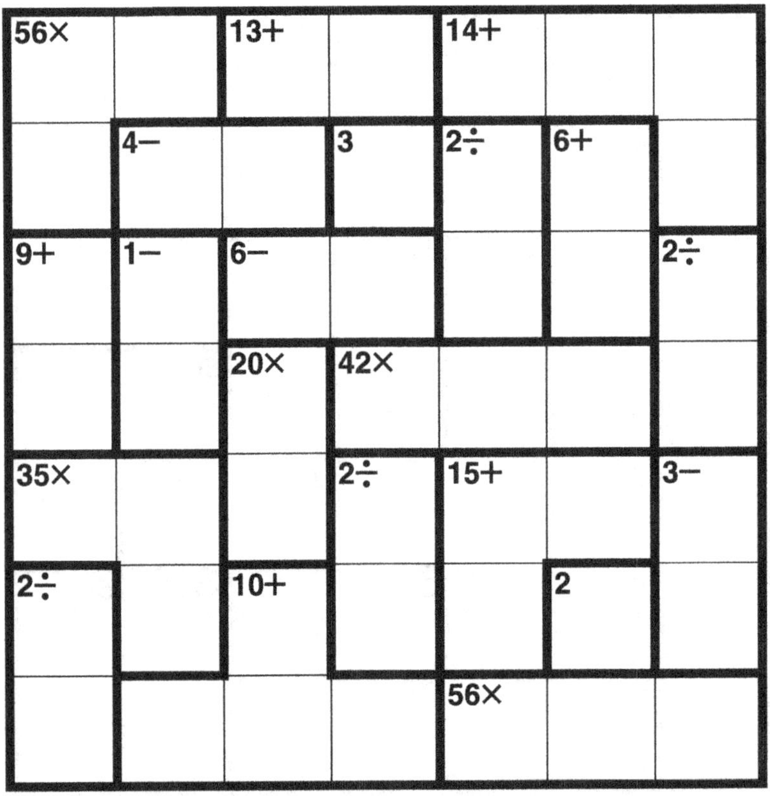

114 + − × ÷

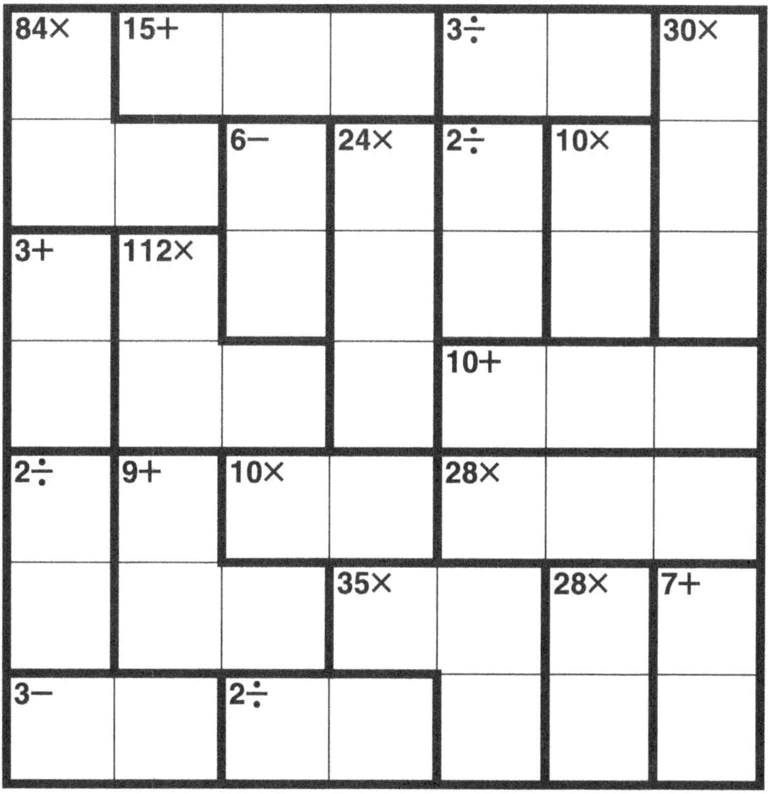

115 + − × ÷

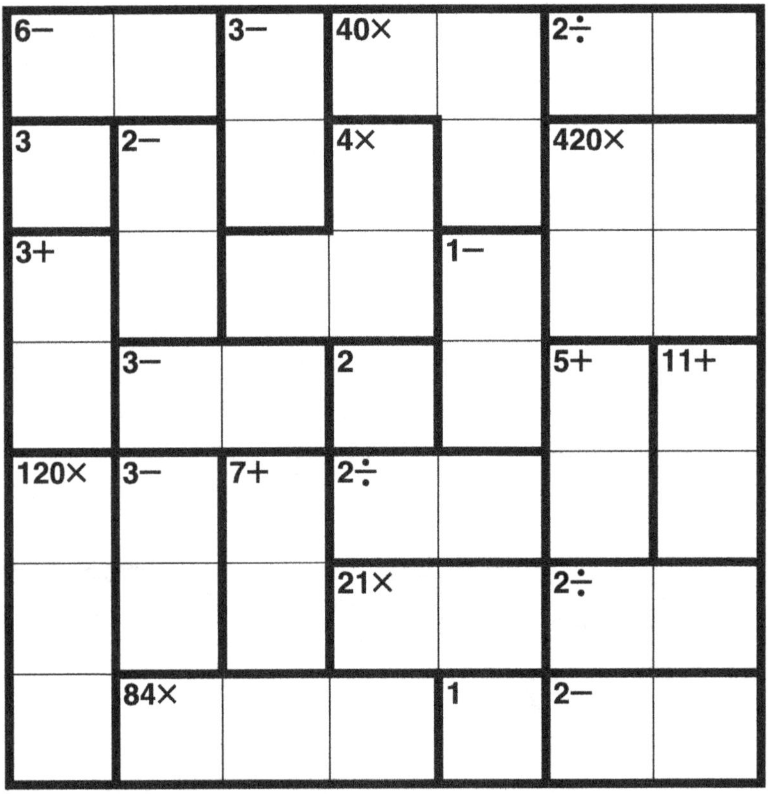

116 + − × ÷

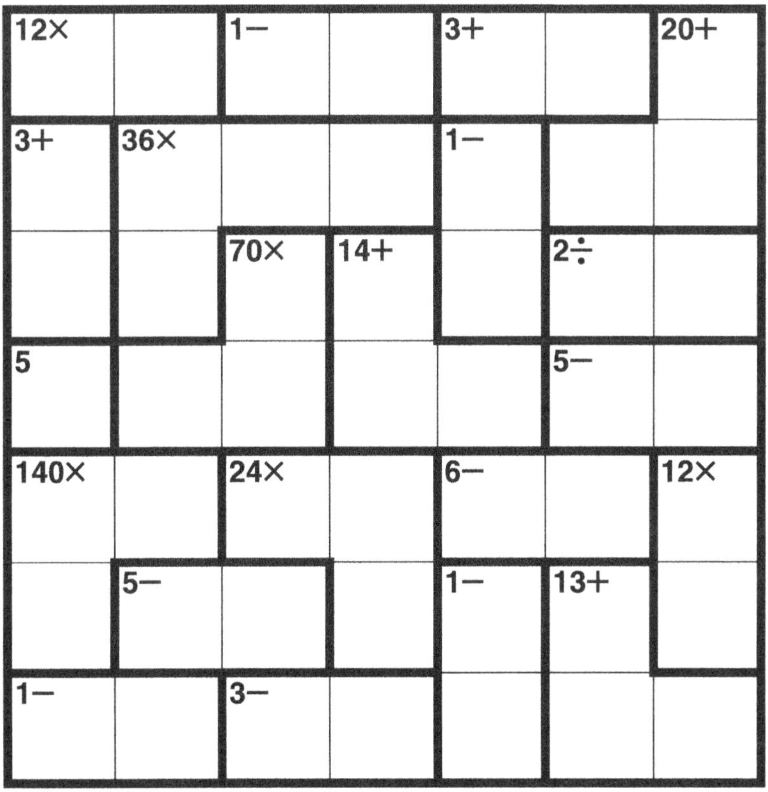

117 + − × ÷

3÷		192×		30×		2−
84×	210×	3+				
			90×			
		13+		42×	16+	
10+		15×			13+	
3÷			3			
	5−		1−			

118 + − × ÷

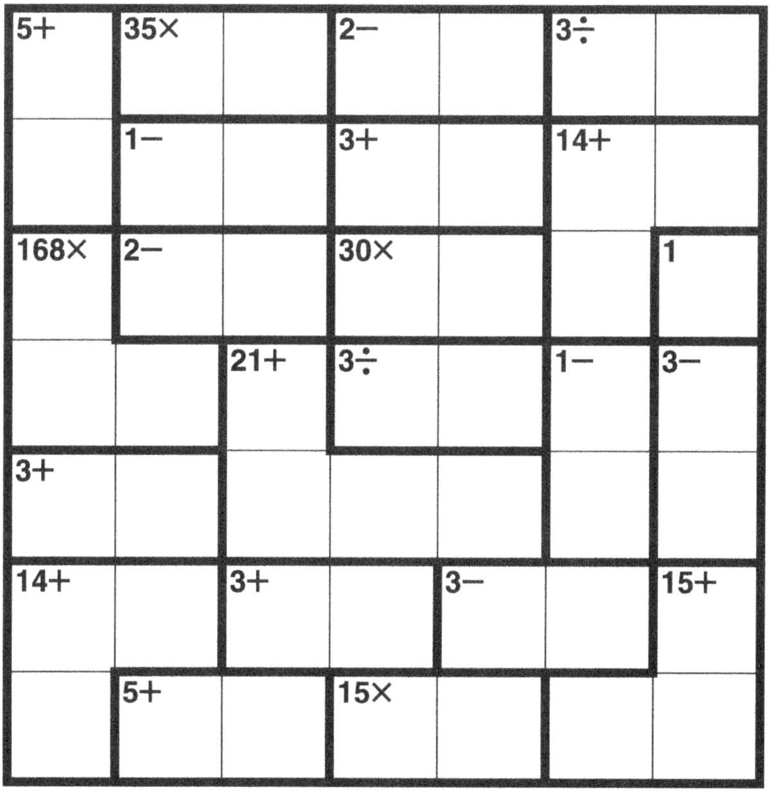

119 + − × ÷

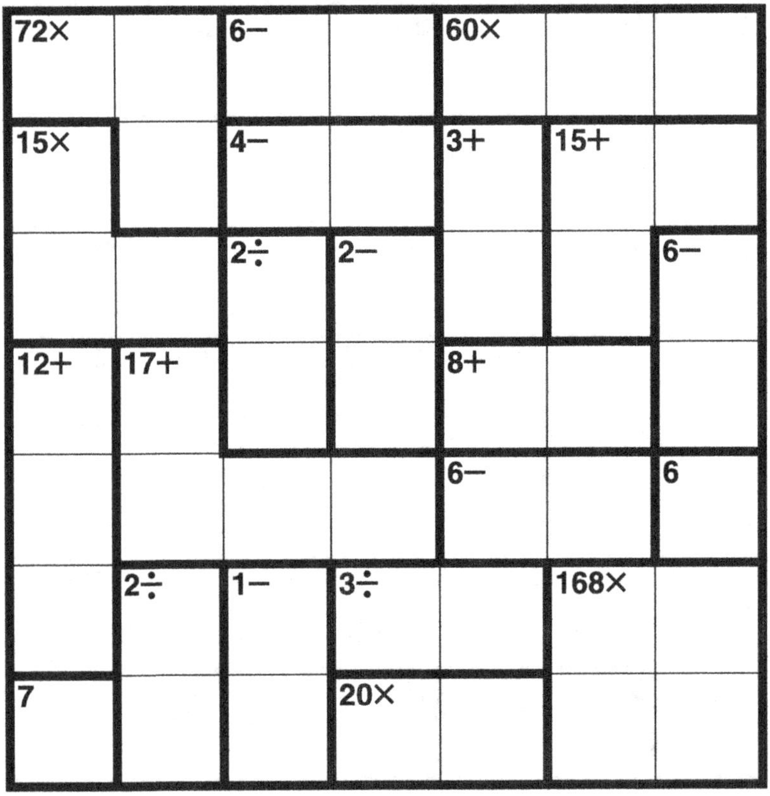

120 + − × ÷

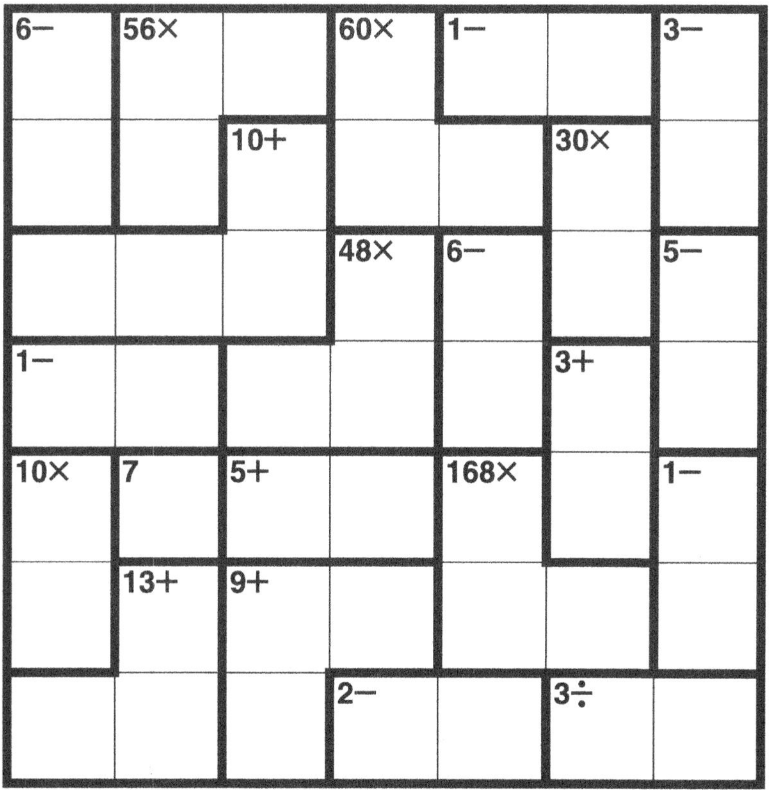

121 + − × ÷

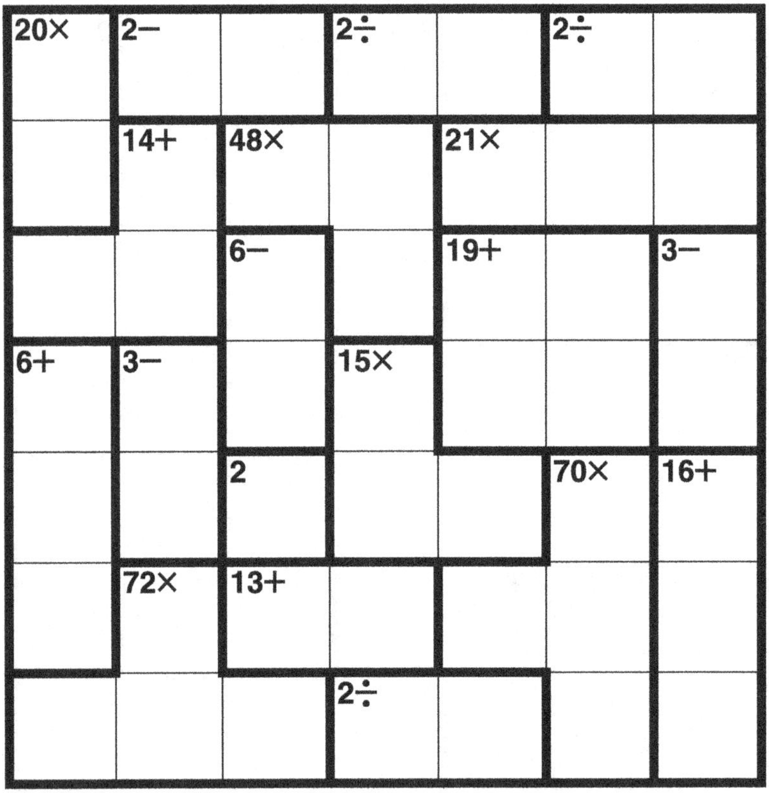

122 + − × ÷

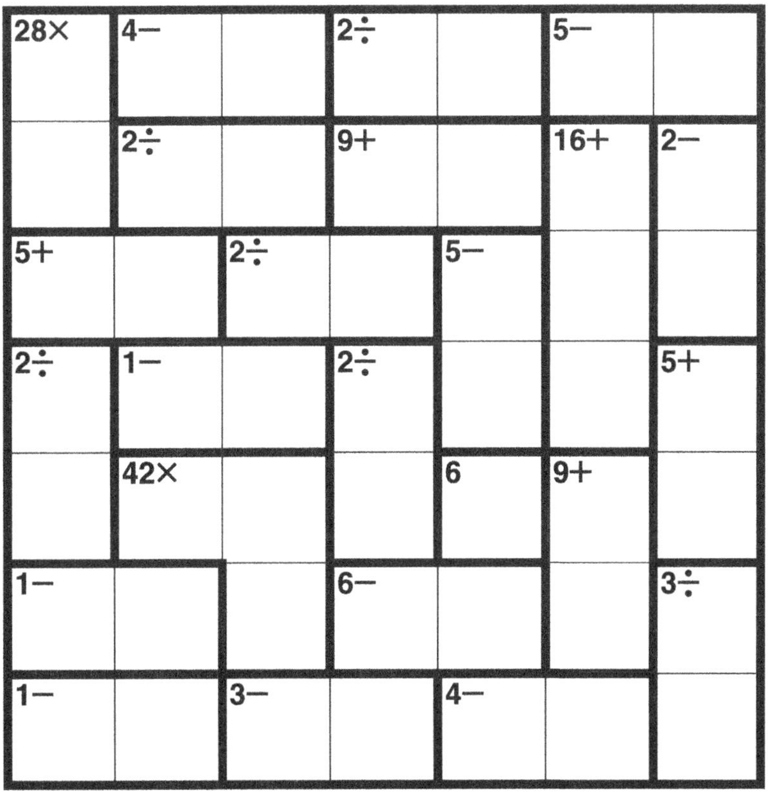

123 + − × ÷

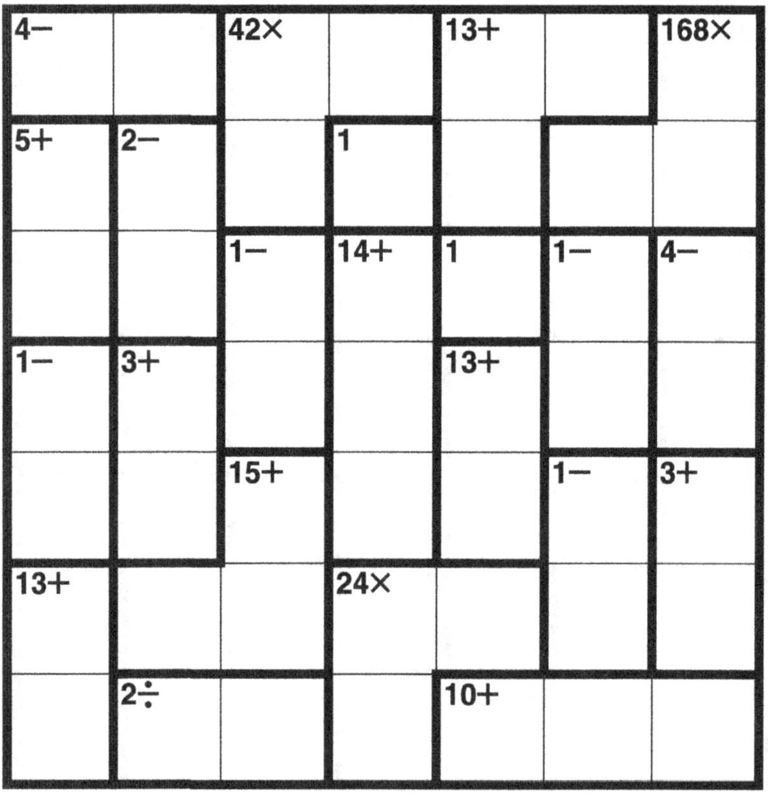

124 + − × ÷

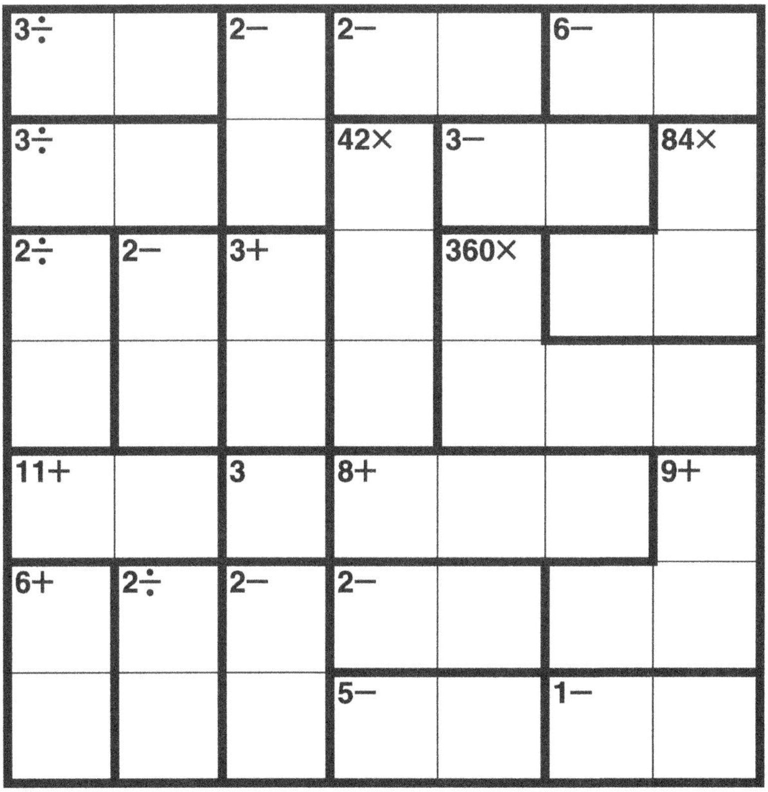

125 + − × ÷

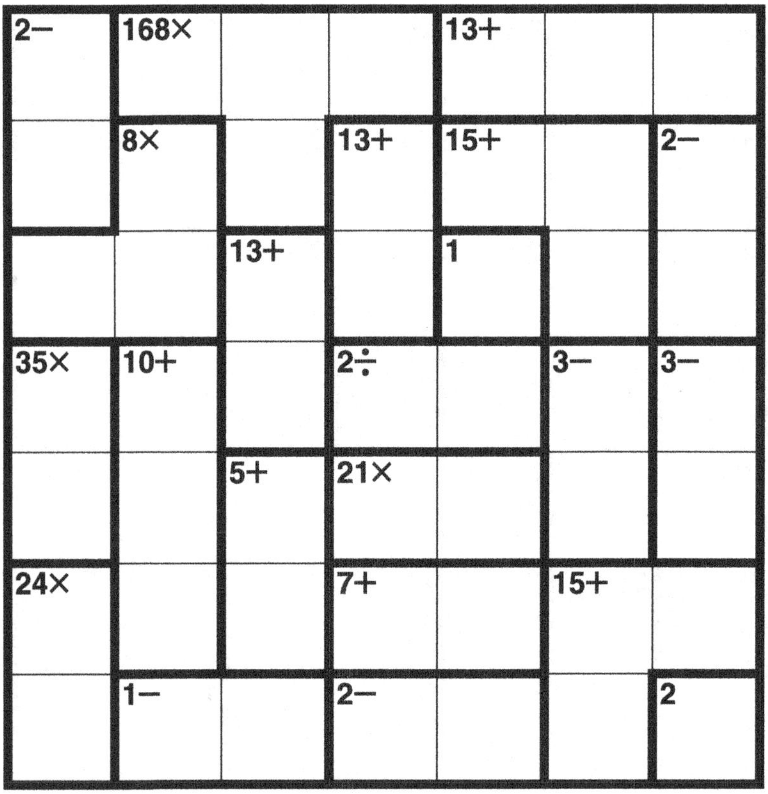

126 + − × ÷

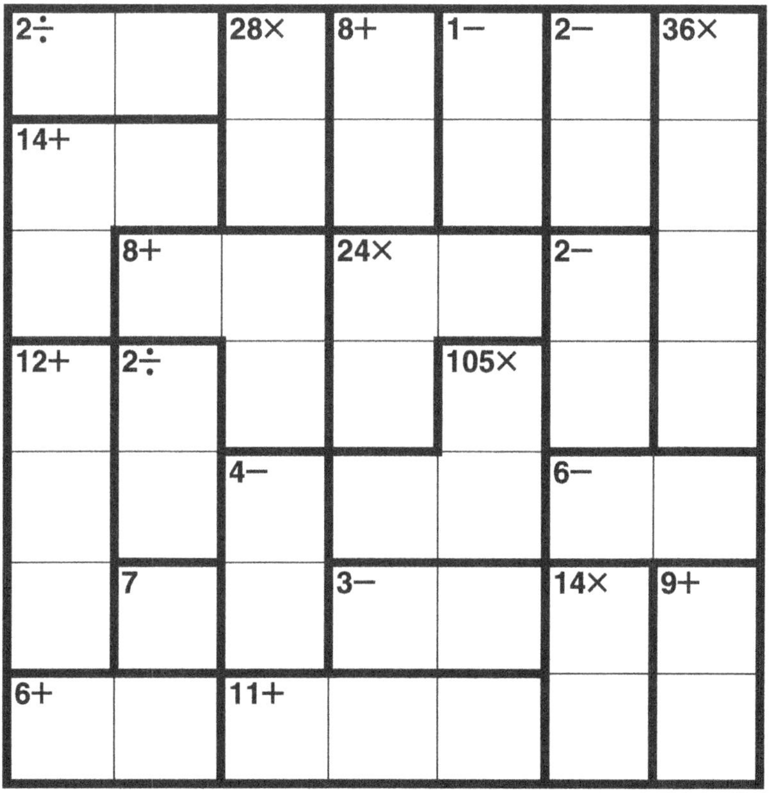

127 + − × ÷

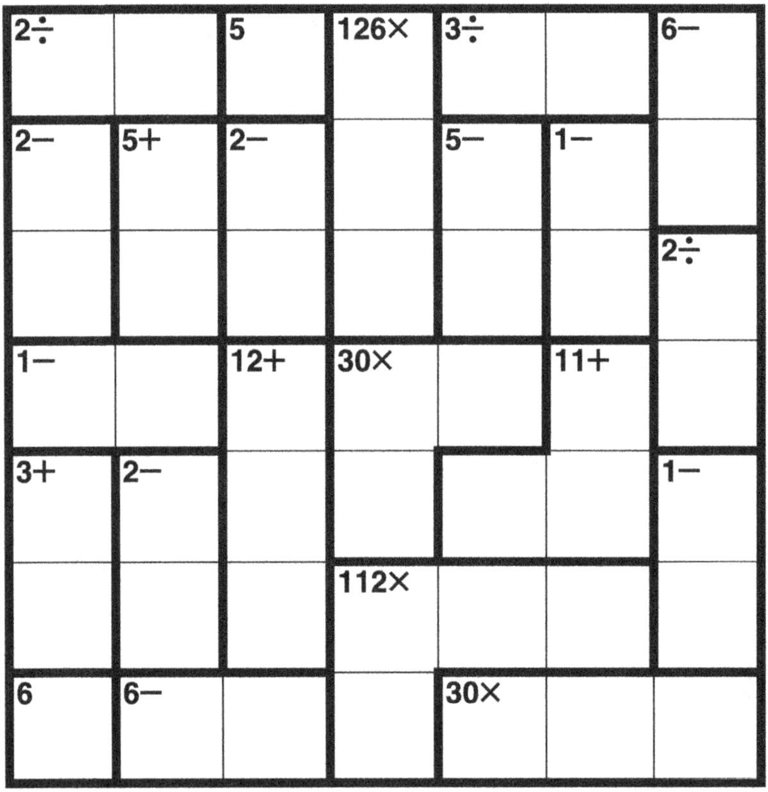

128 + − × ÷

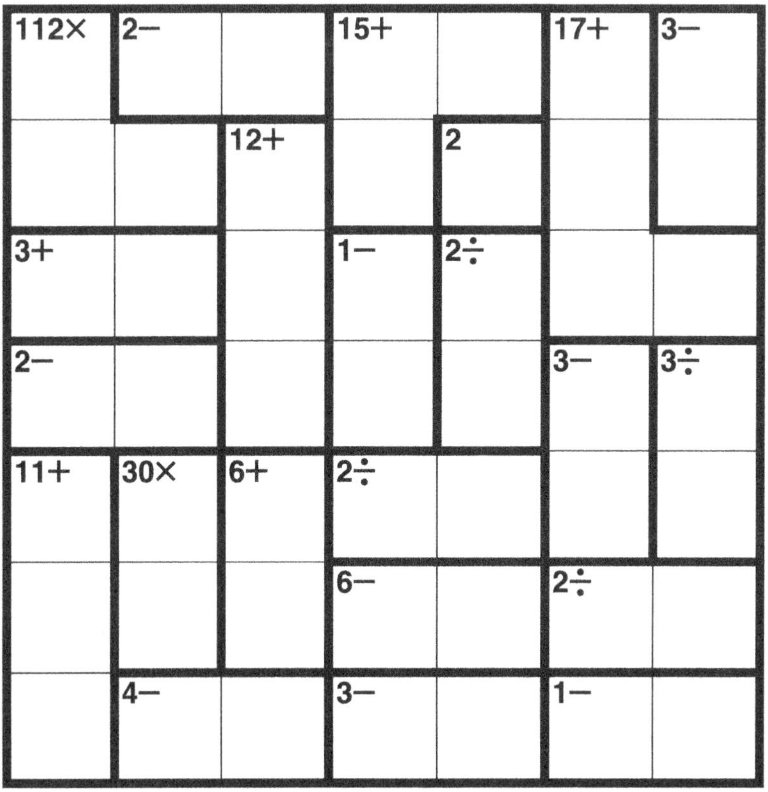

129 + − × ÷

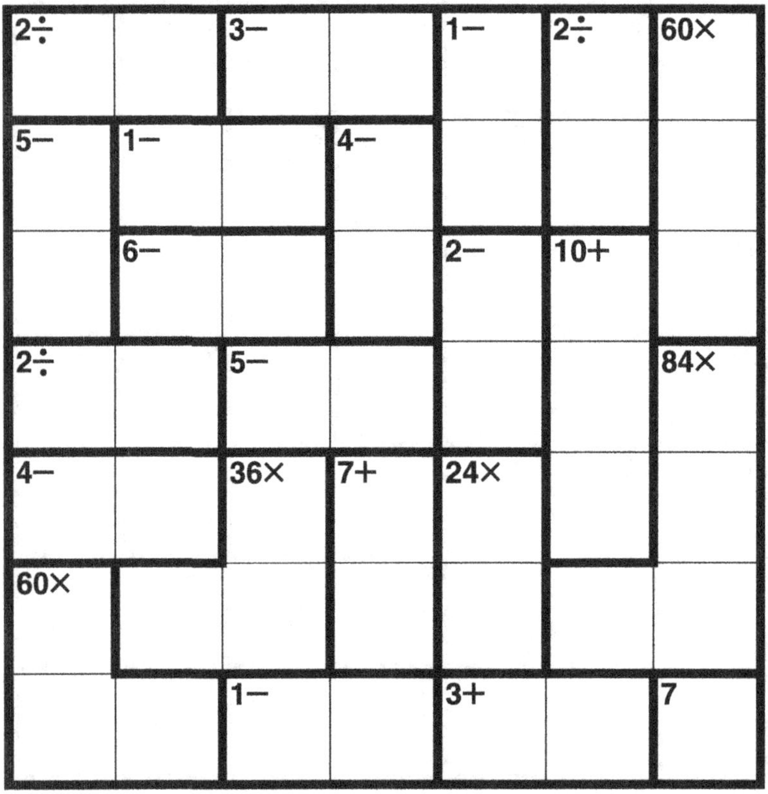

130 + − × ÷

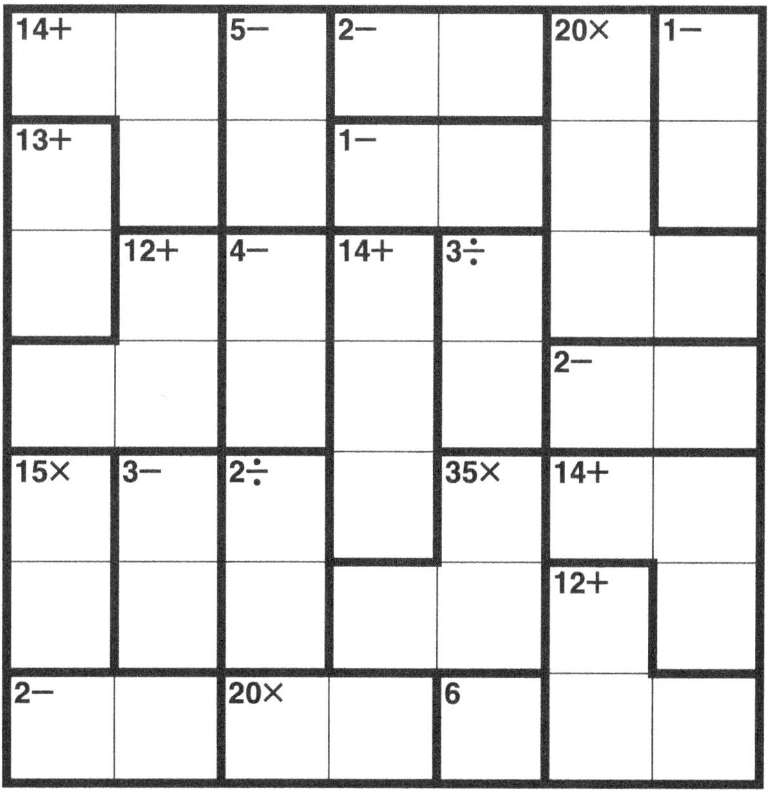

131 + − × ÷

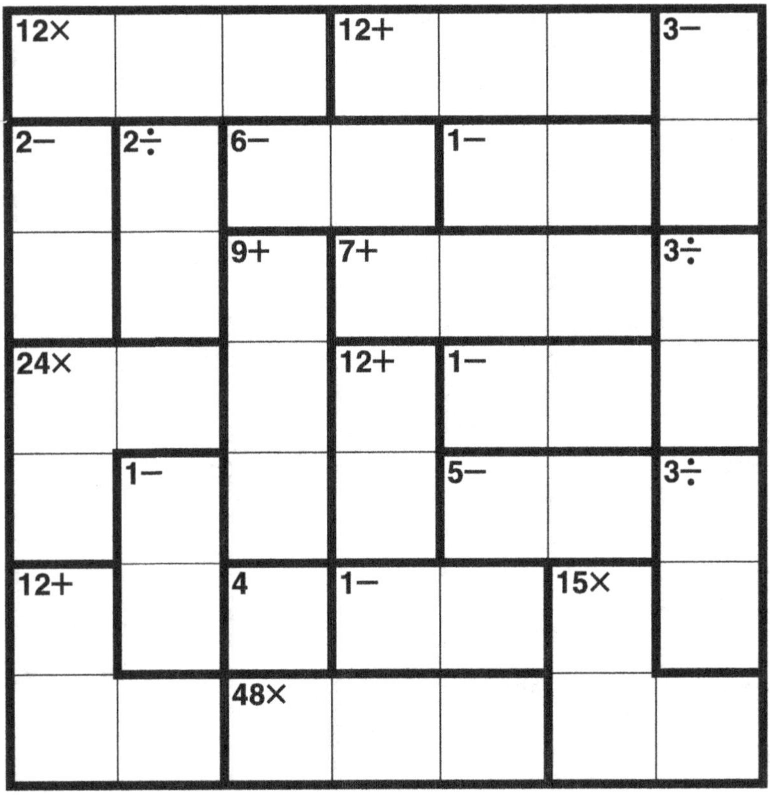

132 + − × ÷

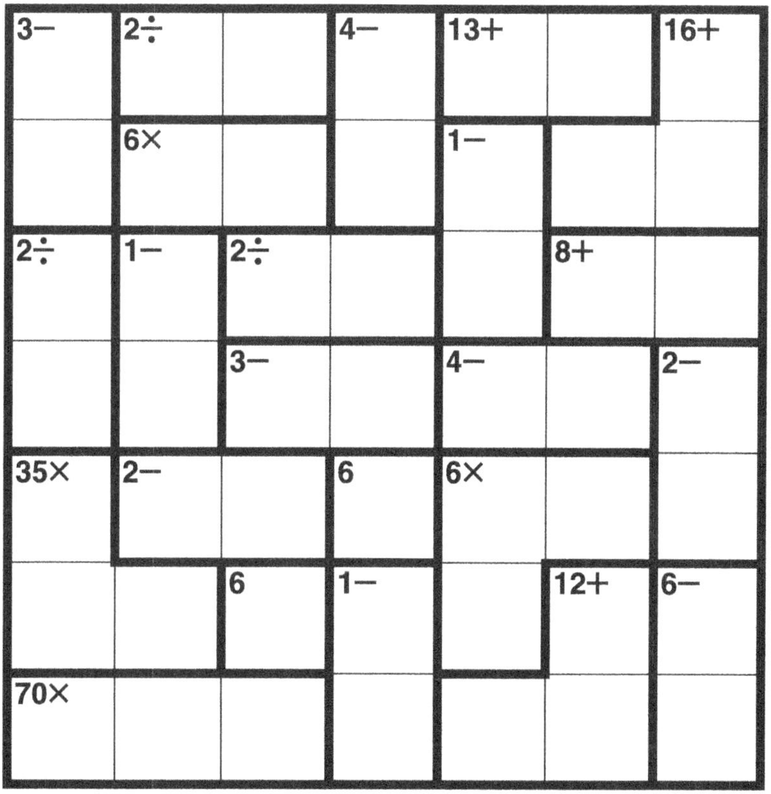

133 + − × ÷

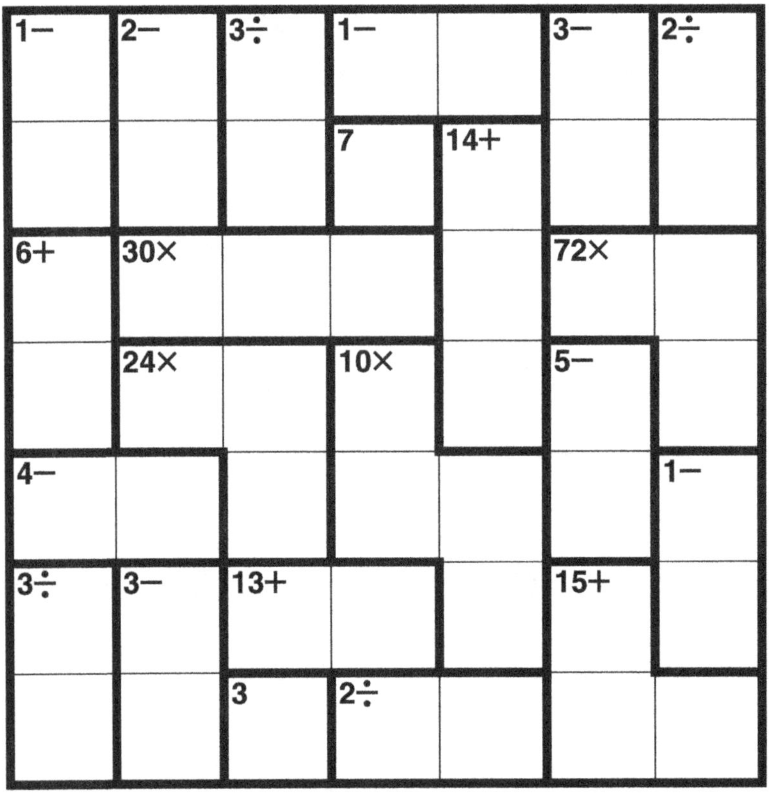

134 + − × ÷

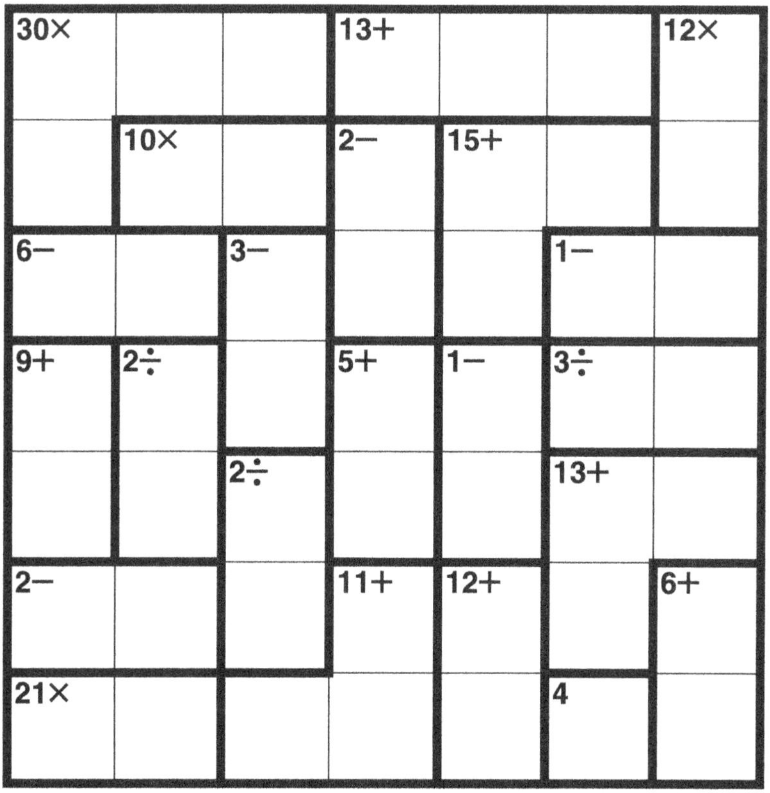

135 + − × ÷

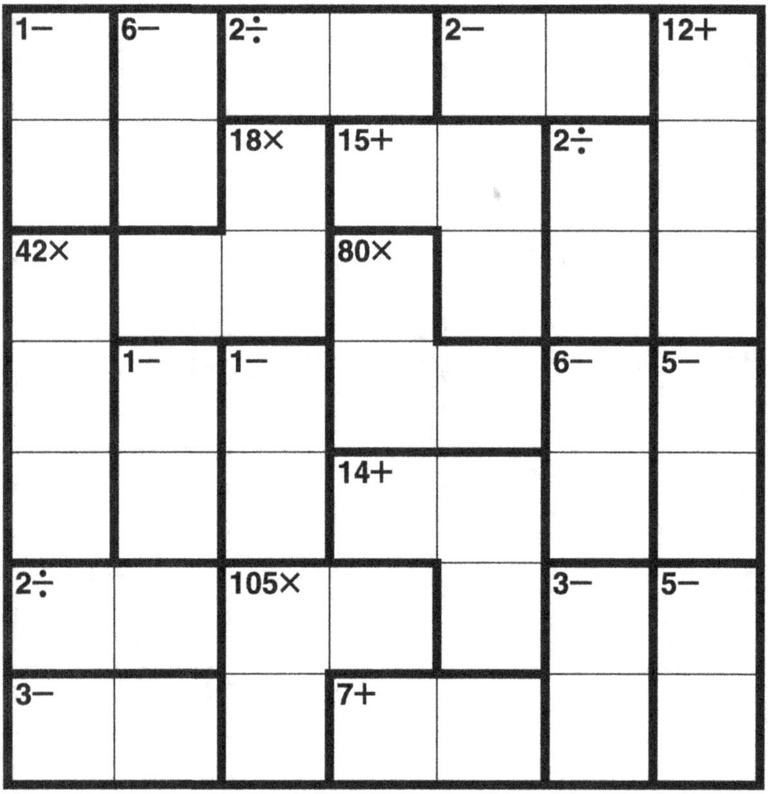

136 + − × ÷

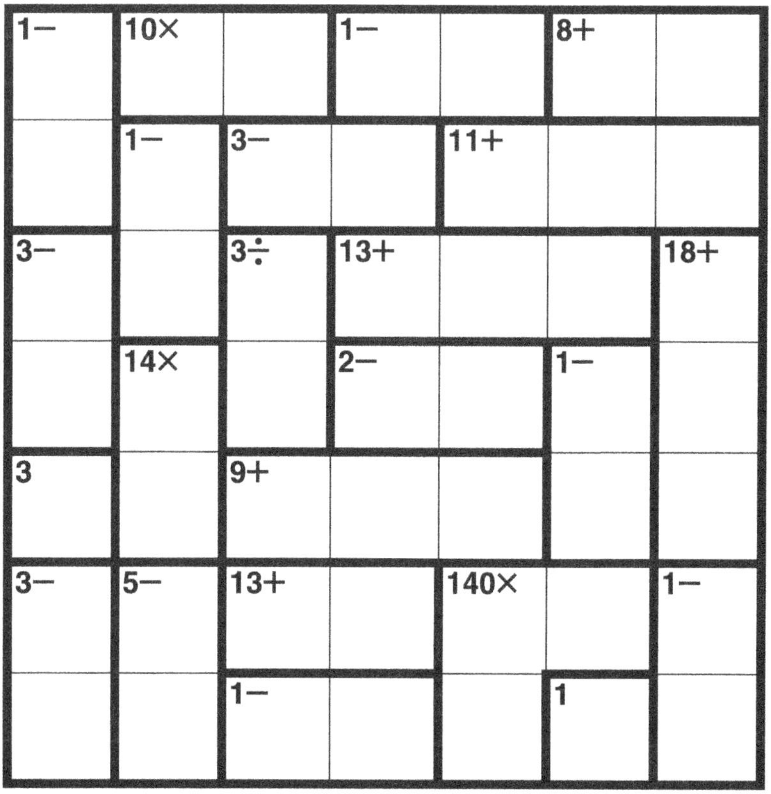

137 + − × ÷

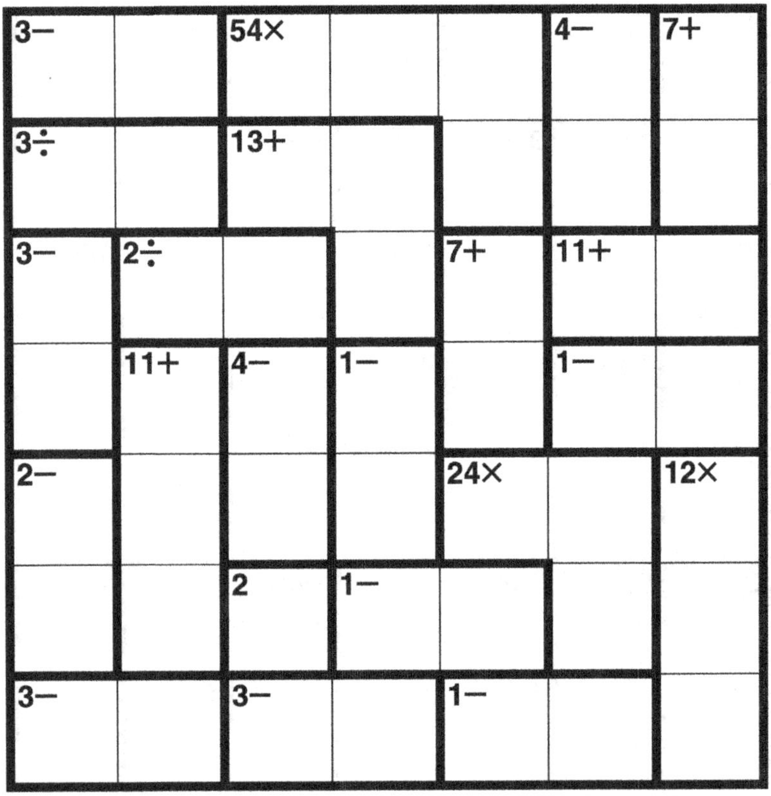

138 + − × ÷

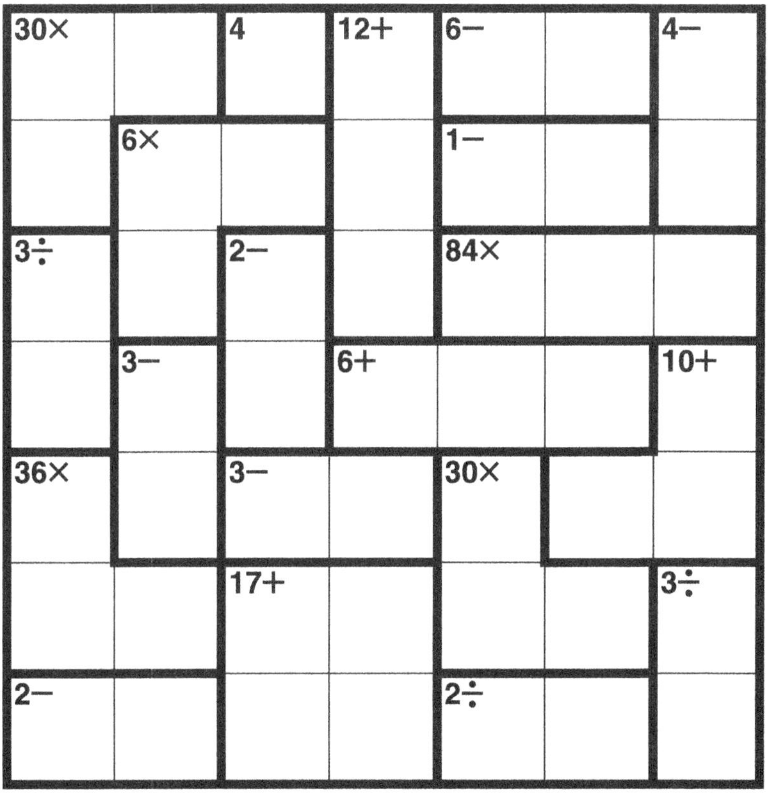

139 + − × ÷

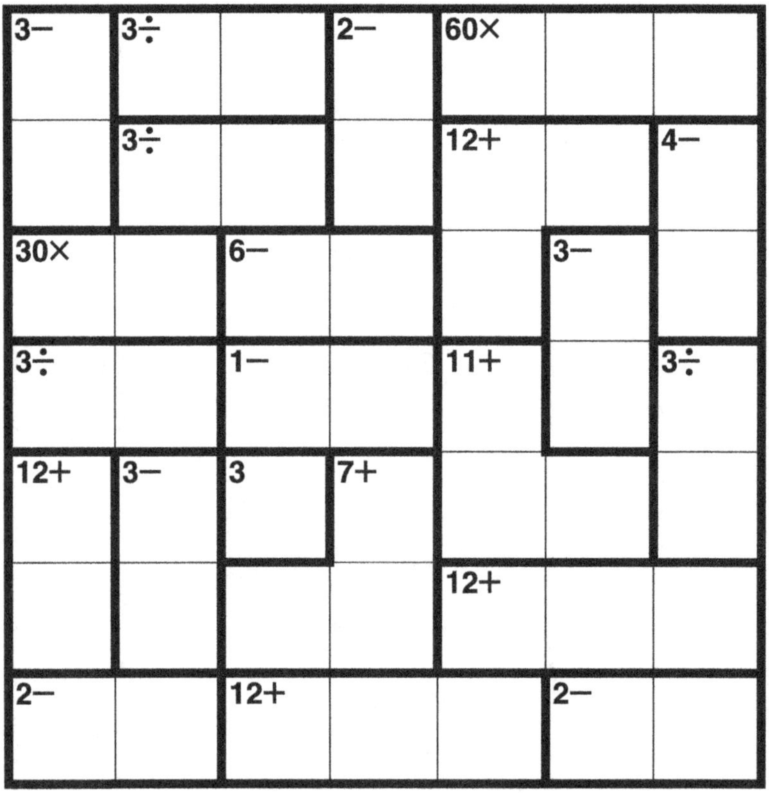

140 + − × ÷

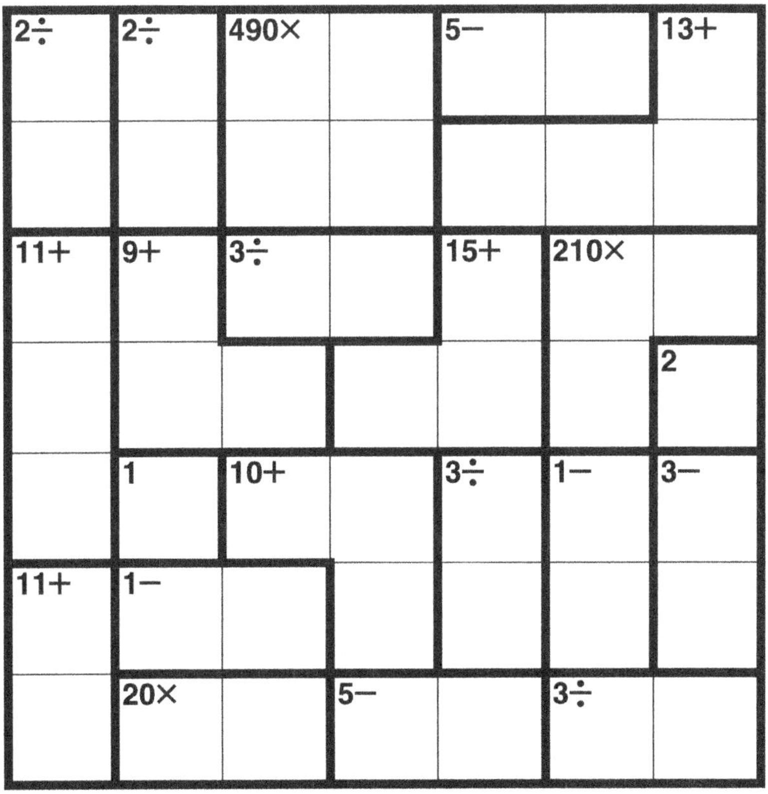

141 + − × ÷

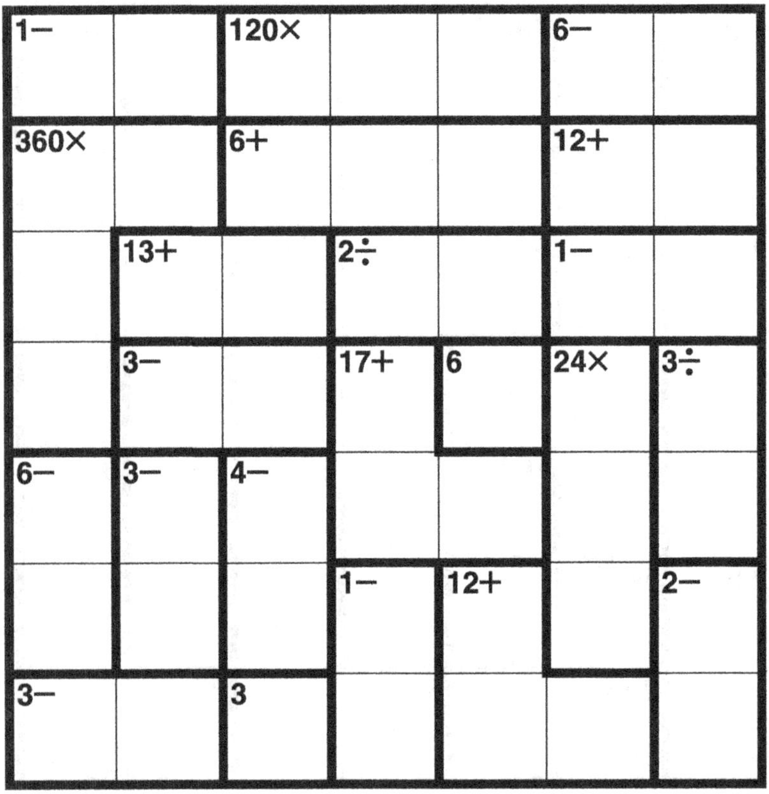

142 + − × ÷

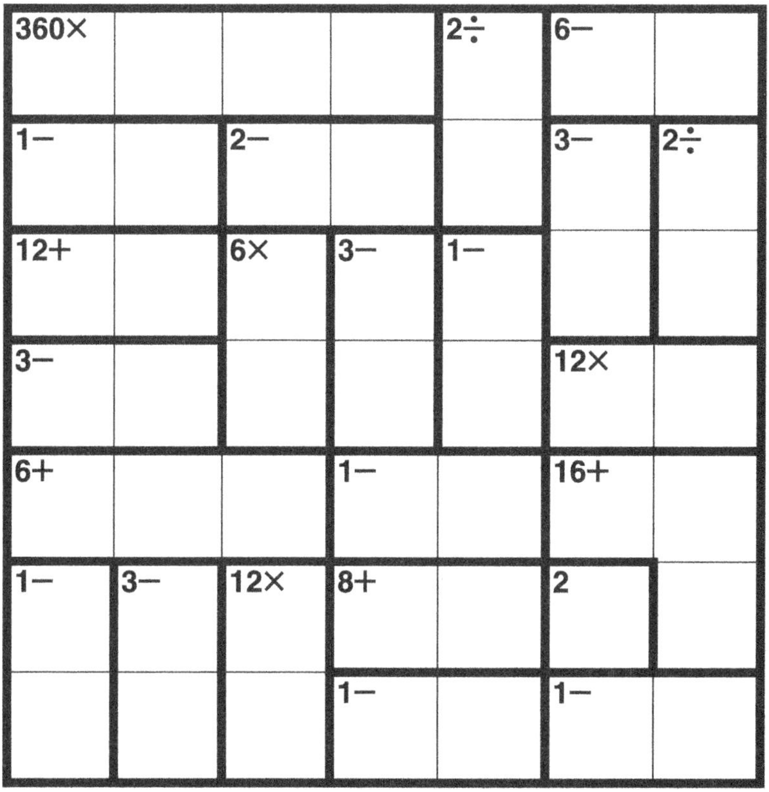

143 + − × ÷

80×	17+		11+		12×	2
			1−			3÷
13+				18+		
8+		3÷				2−
		2−		2−	2−	
1−	2÷		6−			10+
	3−			1−		

144 + − × ÷

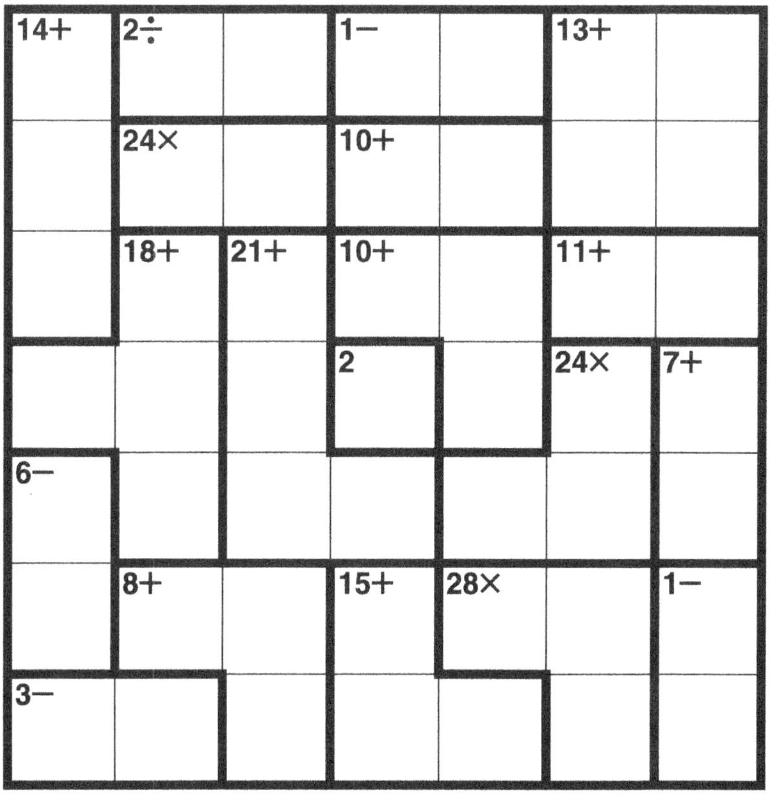

145 + − × ÷

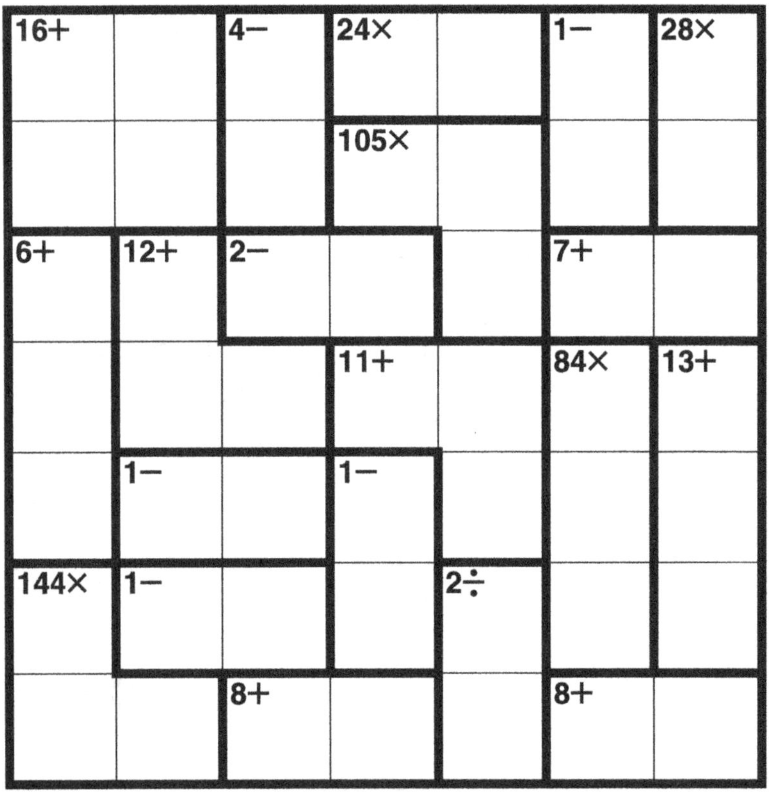

146 + − × ÷

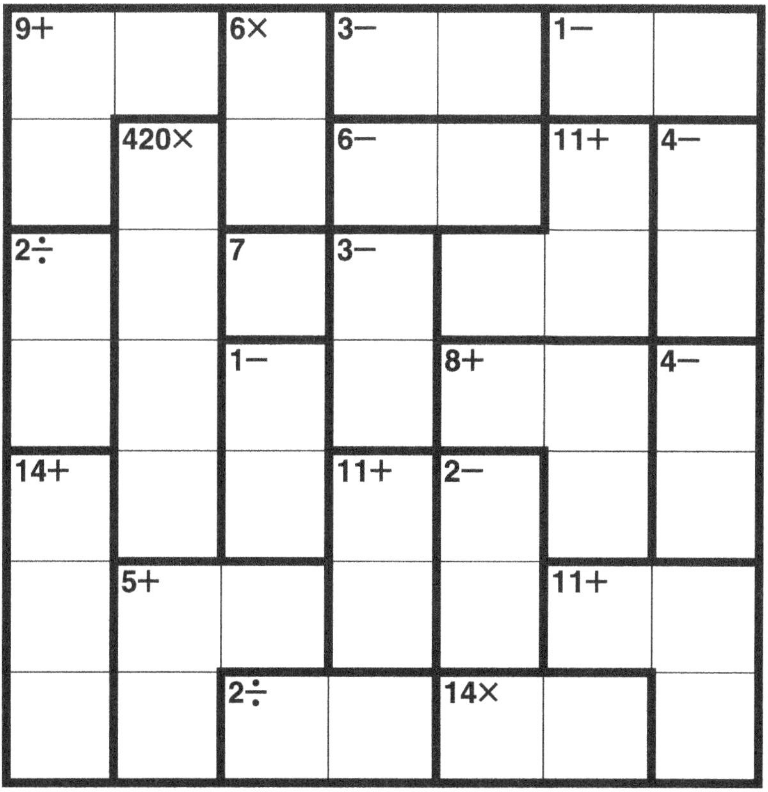

147 + − × ÷

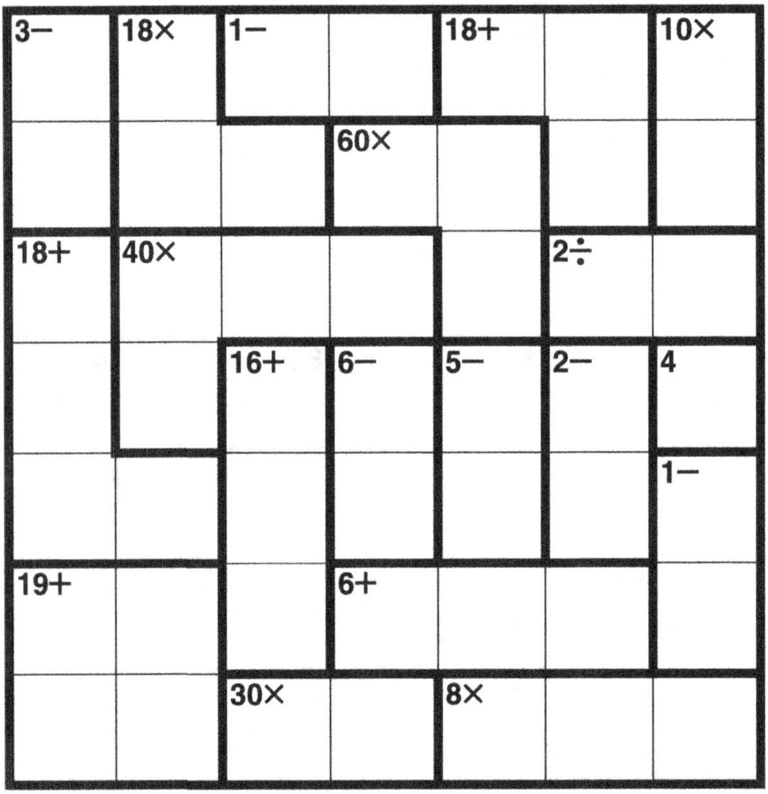

148 + − × ÷

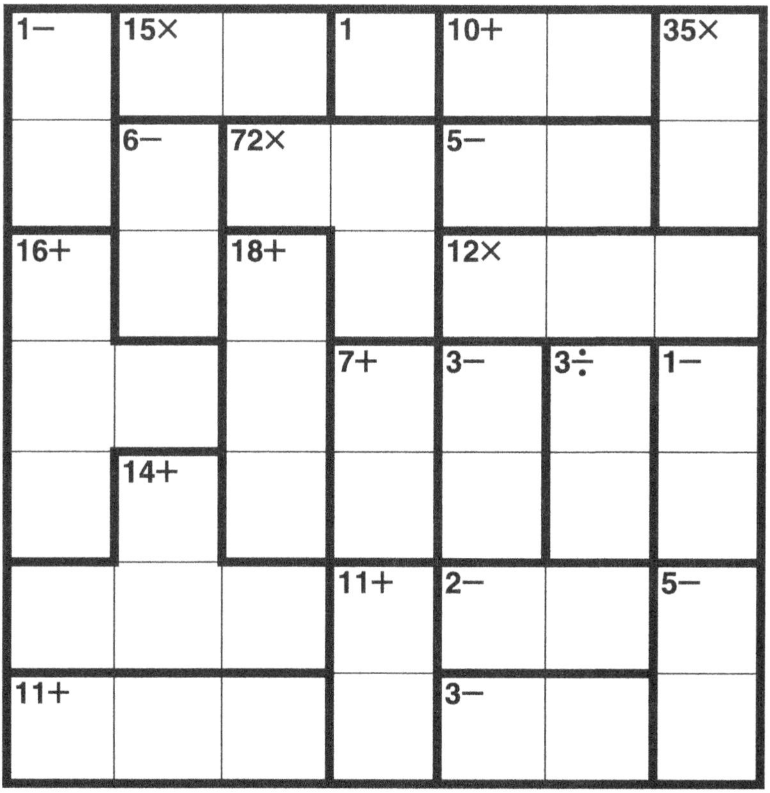

149 + − × ÷

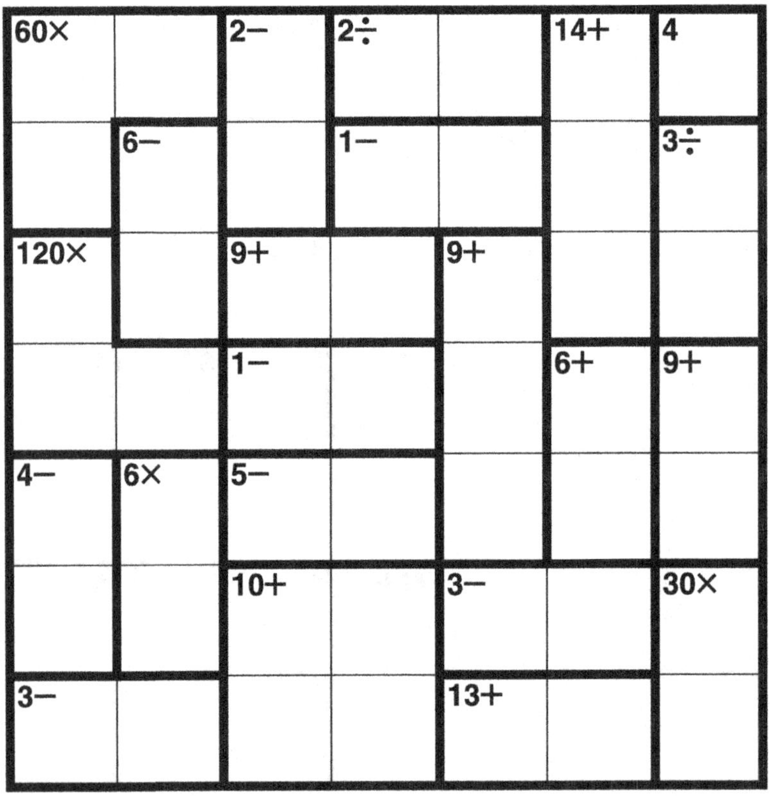

150 + − × ÷

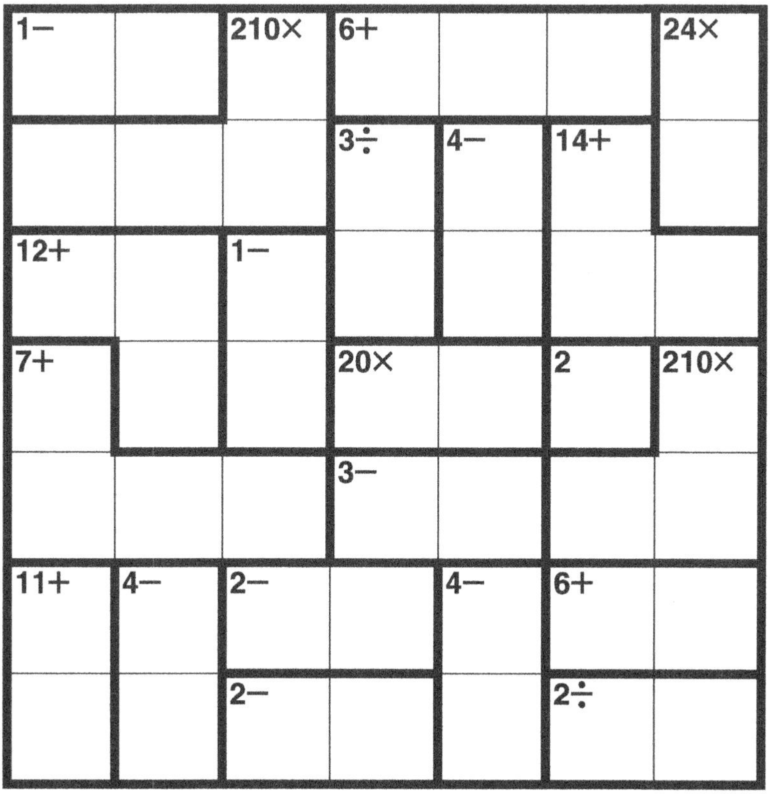

SOLUTIONS

1

9+ 5	2÷ 1	15× 3	2÷ 4	2
4	2	5	4− 1	2− 3
1− 3	4	6× 2	5	1
3− 2	5	1	3	9+ 4
2− 1	3	2÷ 4	2	5

2

5 5	3− 4	1	6× 2	3
4+ 3	1	9+ 4	2− 5	3+ 2
2− 4	2	5	3	1
1− 1	2− 3	2÷ 2	4	9+ 5
2	5	2− 3	1	4

3

3− 2	5	180× 3	4	3− 1
2÷ 1	2	5	3	4
12+ 5	3	4 4	2÷ 1	3− 2
12× 3	4	2÷ 1	2	5
4	1	2	2− 5	3

4

2÷ 1	14+ 5	36× 3	4	2− 2
2	1	5	3	4
2− 5	3	4 4	8+ 2	1
3	2÷ 4	2	15× 1	5
2÷ 4	2	1	5	3

5

1− 3	4	3+ 2	1	6+ 5
2÷ 2	15× 3	300× 5	4	1
4	1	3	5	24× 2
10× 1	5	8× 4	2	3
5	2	1	3 3	4

6

9+ 4	24× 3	2− 1	10× 2	5
5	2	3	10× 1	5+ 4
3 3	4	2	5	1
2÷ 1	1− 5	4	12+ 3	2
2	4− 1	5	4	3

7

2− 1	2− 5	3	2− 4	2
3	4× 2	4− 5	1	1− 4
9+ 4	1	2	5 5	3
5	1− 3	2÷ 4	2	6+ 1
2 2	4	2− 1	3	5

8

4− 1	2− 3	5	6× 2	4 4
5	2 2	3− 4	1	3
1− 4	5	1	5+ 3	2
9+ 2	4	2− 3	5	6+ 1
3	2÷ 1	2	4 4	5

9

9+ 4	5	2− 1	1− 2	3
2− 1	2÷ 2	3	20× 5	2÷ 4
3	1	40× 5	4	2
10× 5	4	2	2− 3	4− 1
2	7+ 3	4	1	5

10

2÷ 2	4	2− 5	3	4− 1
15× 3	1	2÷ 2	4	5
5	2− 3	1− 4	5+ 1	2
3− 1	5	3	2	60× 4
4	2÷ 2	1	5	3

11

60× 4	3	5	2÷ 2	1
3+ 2	4− 5	1	1− 3	4
1	8× 4	15× 3	5	3− 2
8+ 3	1	2− 2	4	5
5	2	8+ 4	1	3

12

1− 3	2	3+ 1	80× 4	5
2÷ 1	12+ 3	2	15× 5	4
2	4	5	1	1− 3
10+ 5	1	1− 4	3	2
4	5 5	3	2÷ 2	1

13

24× 2	4	4+ 3	1	11+ 5
2− 1	3	1− 5	4	2
3	2÷ 1	2	5 5	4
20× 5	2÷ 2	4	1− 3	2− 1
4	4− 5	1	2	3

14

2− 5	3	3+ 2	1	40× 4
3− 1	4	15× 5	7+ 3	2
2÷ 2	1	3	4	5
1− 4	2÷ 2	1	9+ 5	3
3	11+ 5	4	2	1

15

5 5	120× 2	4	2− 3	1
2− 1	5	3	2÷ 4	2
3	8+ 1	3+ 2	3− 5	12+ 4
4	3	1	2	5
2÷ 2	4	6+ 5	1	3

16

15× 3	5	1 1	11+ 4	2
1	4 4	1− 3	2	5
11+ 4	6× 3	2	4− 5	1
5	2	1− 4	2− 1	1− 3
2÷ 2	1	5	3	4

17

15× 3	2÷ 1	5+ 4	3− 2	5
5	2	1	12+ 3	7+ 4
1	11+ 3	5	4	2
2− 2	1− 4	3	5	1
4	5	6× 2	1	3

18

1− 3	12+ 4	2− 1	10× 5	2
2	5	3	1	1− 4
5+ 1	3	2÷ 2	4	5
4	3− 2	5	5+ 3	2− 1
6+ 5	1	4 4	2	3

19

2− 3	1− 4	10× 5	2	1
1	5	6+ 3	2÷ 4	2
1− 2	3	1	1− 5	4
9+ 4	2÷ 1	2	2− 3	5
5	2	3− 4	1	3 3

20

2÷ 4	20× 5	6+ 3	2÷ 1	2
2	4	1	2− 3	1− 5
2− 1	3	2	5	4
10+ 5	1 1	2÷ 4	2	4+ 3
3	2	1− 5	4	1

21

6+ 2	20× 5	3 3	5+ 1	4
3	1	40× 4	5	2
1	4	1− 2	3	2− 5
11+ 5	2	3− 1	4	3
4	2− 3	5	2÷ 2	1

22

30× 3	5	8× 1	4	2
2	1− 3	2÷ 4	6+ 1	4− 5
10+ 5	4	2	3	1
4	4− 1	5	2	1− 3
1	2 2	2− 3	5	4

23

4 4	1− 3	8× 1	2	4− 5
2− 3	2	14+ 5	4	1
1	4	2	2− 5	3
3− 2	4− 5	3	8+ 1	2÷ 4
5	1	4	3	2

24

20× 5	2	7+ 3	4	1 1
2	15× 5	1	3	2÷ 4
2− 3	1	1− 4	5	2
9+ 1	4	30× 5	2	3
4	1− 3	2	6+ 1	5

25

2÷ 2	4	2− 3	2÷ 1	20× 5
9+ 1	1− 3	5	2	4
5	2	1− 4	3	1− 1
3	5 5	3− 1	4	2
3− 4	1	30× 2	5	3

26

2− 1	3	40× 5	4	2
3− 2	2÷ 4	3− 1	15+ 3	5
5	2	4	4− 1	3
1− 3	30× 1	2	5	4
4	5	3	2÷ 2	1

27

120× 2	4	5	4+ 3	1
5+ 4	1− 2	3	4− 1	5
1	3	40× 2	5	4
2− 3	5	3− 1	2÷ 4	1− 2
6+ 5	1	4	2	3

28

20× 5	6+ 2	3	1	12× 4
1	2− 5	2÷ 2	4	3
4	3	1− 5	1− 2	1
6× 2	1	4	2− 3	5
3	5+ 4	1	3− 5	2

29

1− 4	2− 3	1	10× 5	2
3	8+ 2	12+ 5	1	4 4
5	1	4	2÷ 2	9+ 3
2÷ 2	1− 5	3	4	1
1	4	5+ 2	3	5

30

2÷ 2	1	12+ 4	15× 3	8+ 5
2− 1	2− 4	3	5	2
3	2	5	5+ 4	1
1− 5	12× 3	2	1	1− 4
4	5 5	1	2	3

31

2÷ 4	60× 3	5	6+ 1	1− 2
2	10+ 1	4	5	3
5	4	3 3	24× 2	3− 1
2− 1	3− 5	2	3	4
3	2÷ 2	1	4	5 5

32

20× 5	1	1− 2	3	4 4
4	6+ 3	1	2	8+ 5
2− 3	5	60× 4	1	2
1− 1	2	5	8+ 4	3
2÷ 2	4	3	5 5	1

33

60× 3	2− 4	2	4− 5	1
5	1− 1	2÷ 4	2	2− 3
4	2	2− 1	3	5
8+ 2	5	2− 3	3− 1	4
1	3 3	5	2÷ 4	2

34

1− 4	1− 2	3	4− 5	1
3	5+ 4	3− 5	2÷ 1	2
8+ 5	1	2	240× 4	3
2	6+ 5	1	6× 3	4
1	7+ 3	4	2	5

35

4 4	10× 1	5	11+ 3	2÷ 2
1	2	3	5	4
30× 3	1− 5	4	7+ 2	1
2	12× 3	1	4	2− 5
5	4	1− 2	1	3

36

2÷ 4	2÷ 1	2	15× 5	3
2	10+ 4	2− 5	3	1
2− 5	2	3	3− 1	4
3	9+ 5	1	2− 4	2
1	3	4 4	7+ 2	5

37

2− 3	1− 5	4	2÷ 1	2
1	1− 3	2÷ 2	4	20× 5
11+ 5	2	9+ 1	3	4
2	1− 4	3	5	2− 1
4	4− 1	5	2 2	3

38

1− 4	12× 3	1	2÷ 2	8+ 5
5	7+ 2	4	1	3
1	4	2− 3	5	8× 2
2− 3	5	10× 2	1− 4	1
2÷ 2	1	5	3	4

39

2÷ 2	3− 1	4	30× 5	3
1	12+ 4	5	3	2
1− 3	2	1 1	10+ 4	5
1− 4	2− 5	3	2÷ 2	1
5	1− 3	2	1	4 4

40

1− 3	1− 2	10× 1	5	11+ 4
4	3	2	1 1	5
8+ 1	12+ 4	5	3	2
2	2− 5	3	2÷ 4	2− 1
5	3− 1	4	2	3

41

60× 5	3	4	1− 2	1
9+ 2	1− 1	12+ 3	4	5
4	2	8+ 1	2− 5	3
3	5	2	4+ 1	2÷ 4
10+ 1	4	5	3	2

42

4− 1	5	2÷ 4	2	15× 3
2÷ 2	2− 3	5	5+ 4	1
4	1− 2	3	1	5
6+ 3	1	2	20× 5	4
9+ 5	4	6× 1	3	2

43

2 2	40× 4	6+ 5	2− 3	1
2− 3	5	1	2÷ 2	4
1	2	1− 3	4	8+ 5
1− 5	1− 3	4	1	2
4	2÷ 1	2	2− 5	3

44

5 5	15× 3	1	9+ 2	4
1− 2	5	5+ 4	1	3
1	1− 2	3	1− 4	5
1− 4	10+ 1	5	30× 3	2÷ 2
3	4	2	5	1

45

10× 5	1	2	8+ 3	4
1− 3	2	11+ 4	5	1
60× 4	3	1	2	5 5
2÷ 1	5	12× 3	4	1− 2
2	1− 4	5	1	3

46

3− 5	2	3− 1	4	9+ 3
3− 4	1	1− 2	3	5
15× 3	5	4 4	7+ 2	1
2÷ 2	2− 3	5	1	4
1	1− 4	3	3− 5	2

47

40× 2	5	4	4+ 1	15× 3
1− 4	2÷ 2	1 1	3	5
3	1	11+ 2	5	4
4− 5	1− 4	2− 3	1− 2	1
1	3	5	2÷ 4	2

48

4+ 1	3− 5	1− 2	3	1− 4
3	2	20× 4	1	5
3− 4	1	5	11+ 2	3
8+ 5	3	2− 1	4	2
2÷ 2	4	3	4− 5	1

49

1− 3	4	1− 5	2÷ 2	2− 1
3− 5	1− 2	4	1	3
2	3	8× 1	4	60× 5
10+ 1	5	2	3	4
4	2− 1	3	3− 5	2

50

1− 2	3	40× 4	11+ 1	5 5
4− 1	5	2	4	3
9+ 4	2÷ 2	5	3	5+ 1
5	1	2− 3	20× 2	4
1− 3	4	1	5	2

51

2÷ 2	10× 1	5	8+ 4	3
4	2	2− 3	5	1
2− 3	5	3− 4	1	40× 2
4− 5	1− 3	2÷ 1	2	4
1	4	1− 2	3	5

52

2− 3	1	1− 4	5	40× 2
2− 5	3	2÷ 1	2	4
3− 1	4	3− 2	8+ 3	5
8+ 4	2	5	1	2− 3
2	2− 5	3	4	1

53

11+ 5	8× 1	4	2	2− 3
2	4	2− 1	60× 3	5
2÷ 1	2	3	5	3− 4
12× 3	3− 5	2	4	1
4	2− 3	5	2÷ 1	2

54

24× 4	3	2	4− 5	1
8+ 3	5	8+ 1	16× 4	2
3− 5	5+ 1	4	2	2− 3
2	4	3	13+ 1	5
2÷ 1	2	5	3	4

55

60× 3	2÷ 1	2	1− 5	4
5	4	1− 1	2	2− 3
11+ 2	5	10+ 4	3	1
4	2 2	3	5+ 1	3− 5
9+ 1	3	5	4	2

56

5− 1	1− 4	1− 3	2	1− 5	6
6	3	5× 5	1	2 2	2÷ 4
4 4	60× 6	1	120× 5	3÷ 3	2
5	2	6	4	1	2− 3
5+ 3	3+ 1	2	2− 6	4	5
2	5 5	12× 4	3	5− 6	1

57

5− 1	12× 4	3− 5	2	6× 3	11+ 6
6	3	3− 4	3÷ 1	2	5
2− 4	6	1	3	30× 5	2
10+ 3	3÷ 2	6	1− 5	24× 4	1
5	3+ 1	2	4	6	3
2	2− 5	3	6 6	3− 1	4

58

3÷ 3	1	12+ 4	12× 6	3− 2	5
96× 6	4	5	2	3÷ 1	3
4	11+ 6	3	1	5 5	4− 2
3+ 1	5	2÷ 2	4	8+ 3	6
2	10+ 3	6	15× 5	4	1
7+ 5	2	1	3	2− 6	4

59

36× 3	2	11+ 6	5	3− 4	1
2 2	6	1− 4	3	2÷ 1	8+ 5
5− 6	20× 4	5	8× 1	2	3
1	1− 3	2	4	11+ 5	2− 6
1− 5	6+ 1	3÷ 3	2	6	4
4	5	1	6 6	1− 3	2

60

3− 1	3÷ 2	2÷ 6	1− 5	4	6× 3
4	6	3	2 2	2− 5	1
11+ 6	5+ 4	4− 5	1	3	2
5	1	1− 2	3	21+ 6	4
1− 3	12+ 5	1 1	2÷ 4	2	6
2	3	4	5− 6	1	5

61

2÷ 1	2− 6	4	5 5	36× 3	2
2	2− 5	3÷ 1	3	6	8× 4
1− 6	3	9+ 5	4	2	1
5	5+ 2	3	3+ 1	15+ 4	6
7+ 4	5− 1	6	2	5	2− 3
3	2÷ 4	2	5− 6	1	5

62

6× 2	3	14+ 6	5	5+ 1	4
5− 6	360× 4	5	3	2− 2	4− 1
1	6	3	3÷ 2	4	5
9+ 4	2÷ 2	1	6	2− 5	2÷ 3
5	15× 1	2÷ 2	5+ 4	3	6
3	5	4	1	3÷ 6	2

63

60× 3	4	5− 6	1	3− 2	5
5	8+ 3	1	2÷ 4	3÷ 6	2
2÷ 1	11+ 5	4	2	36× 3	5− 6
2	6	30× 5	3	4	1
3÷ 6	2	3	4− 5	1	1− 4
3− 4	1	2	11+ 6	5	3

64

2÷ 3	6	1− 2	3− 5	4 4	24× 1
2÷ 1	4− 5	3	2	6	4
2	1	2− 4	6	2− 5	3
15+ 6	3÷ 3	1	2÷ 4	2	5 5
4	2− 2	11+ 5	3÷ 3	5− 1	6
5	4	6	1	1− 3	2

65

9+ 6	120× 4	5	3+ 2	1	2− 3
2	6	1− 4	3	40× 5	1
1	3 3	11+ 6	5	4	2
3÷ 3	1	2÷ 2	4	3÷ 6	1− 5
1− 4	2− 5	3	1 1	2	6
5	1− 2	1	13+ 6	3	4

66

1− 4	5	5− 1	6	1− 2	3
13+ 5	48× 4	3	2÷ 2	5− 6	1
6	3 3	4	1	9+ 5	2÷ 2
2	3÷ 6	30× 5	3	1	4
3÷ 1	2	6	12× 4	3	11+ 5
3	1− 1	2	1− 5	4	6

67

3− 4	1	30× 6	5	5+ 3	2
1− 5	1− 4	3	3÷ 2	6	10+ 1
6	10+ 3	1 1	8× 4	2	5
3 3	5	3÷ 2	6	1	4
12× 1	2	60× 4	3	5 5	2÷ 6
2	6	5	3− 1	4	3

68

9+ 4	30× 3	5	2	5− 1	6
3	2	96× 6	4	6+ 5	1
1− 1	6+ 5	4	2÷ 6	3	17+ 2
2	1	1− 3	5	6	4
11+ 5	6	2	12× 1	4	3
2− 6	4	3÷ 1	3	3− 2	5

69

6 6	40× 5	2	8+ 4	3	2÷ 1
2− 5	1− 3	4	5− 6	1	2
3	2	10+ 6	1	1− 4	1− 5
2÷ 4	5+ 1	3	2 2	5	6
2	4	1	10+ 5	72× 6	3
5− 1	6	5 5	3	2	4

70

6× 3	1	11+ 6	5	11+ 2	4 4
2	15+ 6	3÷ 1	3	4	5
1− 5	2	4	5− 6	15× 1	2÷ 3
4	3	7+ 2	1	5	6
16+ 6	4	5	12+ 2	3	2÷ 1
1	5	3 3	4	6	2

71

1− 4	5− 1	48× 2	90× 5	3	6
5	6	4	2	3÷ 1	3
1 1	2− 4	6	3	1− 5	10× 2
12+ 2	3	5+ 1	4	6	5
2÷ 3	2	2− 5	2− 6	4	3− 1
6	5	3	3+ 1	2	4

72

11+ 6	1− 4	20× 1	5	6+ 3	2
5	3	4	360× 6	2	1
8× 4	1	2 2	1− 3	5	6
3÷ 1	2	1− 5	4	13+ 6	3
3	180× 5	6	2÷ 2	1	4
2	6	3	3− 1	4	5 5

73

24× 4	2	3	6+ 1	5	15+ 6
11+ 5	6	2÷ 1	2	6× 3	4
8+ 3	1	24× 4	6	2	5
5− 6	4	3− 2	5	8+ 1	1− 3
1	1− 5	6	3	4	2
2 2	2− 3	5	24× 4	6	1

74

8+ 2	1	6 6	60× 4	5	3
5	1− 3	2	1 1	2− 6	4
9+ 4	5	2− 3	2÷ 6	24× 1	2
3÷ 6	2	5	3	4	4− 1
3÷ 1	2− 6	2− 4	2	3	5
3	4	8+ 1	5	2	6 6

75

10× 2	5	3÷ 1	3	2− 4	6
3÷ 3	1	1− 5	4	1− 6	2÷ 2
4− 6	2	10+ 3	1 1	5	4
10+ 1	3	4	3÷ 6	2	2− 5
4	13+ 6	2	5	6+ 1	3
5	2− 4	6	2	3	1 1

76

30× 1	6+ 3	2− 6	1− 5	4	3− 2
6	1	4	11+ 3	2	5
5	2	8+ 1	4	6	3 3
3− 2	5	3	6× 6	1	24× 4
15+ 4	6	5	2÷ 2	2− 3	1
24× 3	4	2	1	5	6

77

12× 4	3	5 5	4− 6	2	5− 1
1	9+ 2	1− 3	1− 5	4	6
6	1	2	6+ 3	1− 5	4
2÷ 2	4	11+ 6	1	3÷ 3	8+ 5
1− 5	6	4	2	1	3
2− 3	5	1	2− 4	6	2 2

78

2− 6	4	2÷ 2	1	15× 5	3
3− 2	3÷ 6	2− 4	9+ 5	3	1
5	2	6	3 3	5+ 1	12+ 4
3÷ 3	11+ 5	1	2− 6	4	2
1	8+ 3	5	4	3÷ 2	6
4	1	1− 3	2	6	5 5

79

5− 1	2− 5	120× 6	4	2÷ 2	1− 3
6	3	5	2÷ 1	4	2
12+ 5	4	3	2	30× 6	11+ 1
2÷ 2	1	1− 4	3	5	6
15+ 3	6	10× 2	5	1	4
4	2	90× 1	6	3	5

80

12+ 3	5 5	2− 2	4	36× 6	1
4	5− 1	30× 3	2	5	6
5	6	2− 1	3	2÷ 2	4
12× 2	15+ 4	6	5	2− 1	3
6	6× 2	1− 4	3÷ 1	3	3− 5
1	3	5	2− 6	4	2

81

1− 2	1	2− 6	6+ 5	1− 3	4
1− 5	2÷ 3	4	1	3÷ 6	2
4	6	1− 2	3	20× 5	1
3÷ 1	7+ 2	3− 3	6	4	11+ 5
3	5	20× 1	4	6× 2	6
2− 6	4	5	2 2	1	3

82

1− 4	6+ 3	1− 5	3÷ 2	6× 6	1
3	2	4	6	1	1− 5
7+ 5	1	3 3	24× 4	2	6
2	7+ 6	1	3	1− 5	4
6× 1	1− 4	11+ 6	5	1− 3	6× 2
6	5	1− 2	1	4	3

83

1− 5	30× 6	9+ 3	4	2	1 1
4	1	5	2÷ 6	3	3÷ 2
1− 3	4	2÷ 2	6+ 1	12+ 5	6
3÷ 6	2− 3	4	2	1	1− 5
2	5	5− 1	3	6	4
2÷ 1	2	6	60× 5	4	3

84

3− 3	6	30× 1	1− 5	4	2 2
10× 1	2	5	6	1− 3	4
5	10+ 3	4	6+ 1	2 2	12+ 6
2− 6	4	3	2	5	1
11+ 4	2÷ 1	2	3	1− 6	5
2	5	6 6	8+ 4	1	3

85

(6) 6	(40×) 2	4	5	(9+) 3	(20×) 1
(144×) 2	6	(2−) 3	1	5	4
4	3	(3÷) 2	6	1	5
(13+) 5	1	6	(360×) 4	(7+) 2	3
1	(12+) 4	5	3	6	2
3	5	(3+) 1	2	(2−) 4	6

86

(10+) 5	4	(3) 3	(12×) 6	1	2
1	(2÷) 3	6	(36×) 2	(1−) 5	4
(4) 4	(10×) 2	5	3	6	(5−) 1
(5+) 2	(5−) 1	(2÷) 4	(2−) 5	3	6
3	6	2	(9+) 1	4	(10+) 5
(12+) 6	5	1	4	2	3

87

(72×) 4	(3÷) 6	2	(6+) 3	(30×) 5	(3−) 1
6	3	(12+) 5	2	1	4
(2−) 5	4	3	1	6	(3÷) 2
3	(8+) 5	1	(2−) 4	2	6
(1) 1	2	(2−) 6	(1−) 5	(1−) 4	3
(2÷) 2	1	4	6	(15×) 3	5

88

(12×) 1	2	6	(14+) 4	(13+) 3	5
(10+) 3	1	4	6	5	(6×) 2
6	(5+) 4	1	(3−) 5	(2÷) 2	3
(9+) 4	3	(2−) 5	2	1	(24×) 6
2	(16+) 5	3	(5−) 1	6	4
5	6	(9+) 2	3	4	1

89

5− 1	6	11+ 4	10× 2	5	11+ 3
2− 2	4	6	1	3	5
9+ 5	1	1− 2	3	96× 4	6
3	2÷ 2	1	60× 5	6	4
6 6	180× 3	5	4	2	1
1− 4	5	3	12× 6	1	2

90

120× 5	4	6	6+ 2	3	1
11+ 2	3	5− 1	6	3− 4	10× 5
6	2− 5	3	1− 4	1	2
2÷ 1	2	2− 4	3	30× 5	6
3− 4	1	2	60× 5	6	7+ 3
3− 3	6	4− 5	1	2	4

91

12× 4	2− 3	5	10+ 2	5− 6	1 1
3	2÷ 4	2	6	1	13+ 5
6+ 5	2	3÷ 1	3	7+ 4	6
1	1− 6	4 4	4− 5	3	2
3÷ 6	5	18× 3	1	3− 2	1− 4
2	1	6	4 4	5	3

92

60× 6	6× 1	3	2	13+ 5	4
2	11+ 3	5− 1	6	4	5 5
5	2	6	24× 4	3÷ 1	3
12× 4	1− 6	5 5	3	2	9+ 1
3	5	8× 4	1	6	2
5+ 1	4	2	5 5	2÷ 3	6

93

13+ 4	5	3 3	12× 2	5− 1	6
2− 5	4	2	3	6 6	1− 1
3	1− 6	5	3− 1	4	2
2÷ 1	2	2− 4	6	1− 3	1− 5
11+ 6	3	6× 1	1− 5	2	4
2	1	6	4	8+ 5	3

94

15× 5	3	1	4− 2	6	5+ 4
15+ 4	6	2− 3	5	2÷ 2	1
2÷ 3	5	5− 6	1	4	3− 2
6	2÷ 4	2	2÷ 3	4− 1	5
4× 2	20× 1	4	6	5	2÷ 3
1	2	5	1− 4	3	6

95

2− 4	3− 6	10× 5	2	3÷ 3	1− 1
6	3	1− 4	5	1	2
10+ 5	4	2÷ 2	1	2÷ 6	3
1	10× 2	1− 3	4	60× 5	1− 6
6× 2	1	5− 6	3	4	5
3	5	1	48× 6	2	4

96

2− 4	6	24× 1	3	13+ 5	2
8+ 1	1− 5	2	4	6	2− 3
5	4	2÷ 3	6	7+ 2	1
2	2− 3	5	1	4	120× 6
3− 6	5+ 1	4	6+ 2	3	5
3	60× 2	6	5	1	4

97

40× 4	3÷ 3	1	60× 5	6	2
2	9+ 1	2− 5	3	2− 4	6
5	6	2	1− 4	3	9+ 1
48× 6	4	6+ 3	1	2	5
3÷ 1	2	2− 4	6	20× 5	3
3	13+ 5	6	2	1	4

98

2÷ 4	1− 6	5	2÷ 3	2− 1	1− 2
2	2− 5	48× 4	6	3	1
24× 6	3	2	11+ 1	5	4 4
1	4	6	40× 5	2	3
15× 3	2	1	4	48× 6	1− 5
5	3÷ 1	3	2	4	6

99

18× 3	1− 5	2÷ 4	2	9+ 1	1− 6
1	4	2− 3	6	2	5
6	3÷ 2	5	12+ 4	3÷ 3	1
40× 4	6	5− 1	3	3− 5	2
2	2− 1	6	5	12× 4	3
5	3	2÷ 2	1	2− 6	4

100

18+ 2	5− 6	1	2÷ 3	60× 5	4
5	1	4	6	7+ 2	3
6	40× 4	5	2	3	1
12× 4	10+ 3	2	7+ 1	6	3− 5
3	5	72× 6	4	10+ 1	2
1	2	3	5	4	6 6

101

100× 4	5	12× 1	6	8+ 2	3
5	2− 6	4	2	3	24× 1
5− 1	12+ 2	3	4− 5	4	6
6	8+ 3	2	1	20× 5	4
5+ 3	1	5	2− 4	6	10× 2
2	4	2÷ 6	3	1	5

102

3− 3	6	25× 5	1	3− 4	2÷ 2
15+ 6	1− 3	2÷ 2	5	1	4
5	2	1	2÷ 4	14+ 6	3
4	3÷ 1	3	2	5	12+ 6
2÷ 1	2− 4	6	2÷ 3	1− 2	5
2	1− 5	4	6	3	1

103

2÷ 1	2	5− 6	16+ 5	12× 4	3− 3
10+ 4	5	1	2	3	6
10+ 2	1	4 4	3	6	4− 5
5	72× 6	3	4	80× 2	1
3	60× 4	5	9+ 6	1	2
6 6	3	2	1	5	4

104

60× 1	11+ 5	6	7+ 3	4	7+ 2
3	15+ 6	1	2− 4	2	5
4	3	5	3÷ 2	6	5+ 1
5	8+ 2	4 4	10+ 6	1	3
2	4	3	1	19+ 5	6
9+ 6	1	2	5	3	4 4

105

10+ 3	1− 4	5	3÷ 6	2	5− 1
4	2	15× 3	5	1	6
1	90× 5	6	72× 3	4	3− 2
60× 2	3	3− 1	4	6	5
6	3− 1	4	3− 2	5	1− 3
5	3÷ 6	2	2− 1	3	4

106

2÷ 2	11+ 3	20× 4	5	1	15+ 6
4	6	2÷ 1	2	30× 3	5
2÷ 3	2	30× 6	1	5	4
6	15× 1	5	14+ 4	2	1− 3
4− 1	5	3	6	4	2
5	2÷ 4	2	10+ 3	6	1

107

2÷ 6	3	1− 4	5	10× 1	2
3 3	4− 2	2− 6	4	5	1
4− 5	6	1− 2	3− 1	4	8+ 3
1	20× 4	3	3÷ 2	6 6	5
2÷ 2	5	1	6	1− 3	2− 4
4	1 1	2− 5	3	2	6

108

10× 5	6+ 3	2	1	11+ 6	4
2	8+ 5	3	10+ 6	4	1
5+ 1	2− 6	4	10+ 5	2	3− 3
4	1 1	3− 5	2	3	6
2÷ 6	2− 4	6× 1	1− 3	7+ 5	2
3	2	6	4	6+ 1	5

109

6× 3	2	15+ 4	6	5	6× 1
11+ 4	5	2	3÷ 3	1	6
90× 1	2− 6	3 3	3− 5	2	1− 4
5	4	2÷ 1	2	48× 6	3
6	3	12+ 5	1	4	2
2÷ 2	1	6	4 4	2− 3	5

110

15× 5	1	3	8+ 2	14+ 6	4
2 2	14+ 3	5	1	4	7+ 6
6	5	15+ 2	1− 4	3	1
2÷ 1	2	4	6	75× 5	3
2− 4	6	6× 1	3	10+ 2	5
1− 3	4	6	5	1	2

111

168× 2	4	3	3÷ 6	2− 1	35× 5	7
2÷ 6	1 1	7	2	3	480× 4	5
3	3+ 2	1	6− 7	3− 5	6	4
1− 4	5	2− 6	1	2	21× 7	3
3÷ 1	3	4	2− 5	7	3÷ 2	6
2− 7	30× 6	5	1− 3	3− 4	1	2 2
5	5− 7	2	4	6 6	4+ 3	1

112

60× 4	3	2− 6	13+ 7	30× 1	5	2
5	8× 1	4	6	3	5− 2	7
5− 1	4	6− 7	3− 2	12+ 5	2÷ 3	6
6	2	1	5	7	60× 4	3
42× 7	13+ 6	10+ 2	3	3− 4	1	5
3	7	5	3+ 1	2	42× 6	5+ 4
2	8+ 5	3	10+ 4	6	7	1

113

56× 2	4	13+ 6	7	14+ 5	3	1
7	4− 6	2	3 3	2÷ 4	6+ 1	5
9+ 4	1− 3	6− 7	1	2	5	2÷ 6
5	2	20× 4	42× 6	1	7	3
35× 1	7	5	2÷ 2	15+ 3	6	3− 4
2÷ 3	5	10+ 1	4	6	2 2	7
6	1	3	5	56× 7	4	2

114

84× 4	15+ 5	3	7	3÷ 2	6	30× 1
7	3	6− 1	24× 4	2÷ 6	10× 2	5
3+ 2	112× 4	7	1	3	5	6
1	7	4	6	10+ 5	3	2
2÷ 3	9+ 6	10× 5	2	28× 4	1	7
6	1	2	35× 5	7	28× 4	7+ 3
3− 5	2	2÷ 6	3	1	7	4

115

6− 7	1	3− 4	40× 5	2	2÷ 3	6
3 3	2− 5	7	4× 1	4	420× 6	2
3+ 2	3	1	4	1− 6	7	5
1	3− 6	3	2 2	5	5+ 4	11+ 7
120× 5	3− 7	7+ 2	2÷ 6	3	1	4
6	4	5	21× 3	7	2÷ 2	1
4	84× 2	6	7	1 1	2− 5	3

116

12× 3	4	1− 6	5	3+ 1	2	20+ 7
3+ 2	36× 1	3	4	1− 5	7	6
1	3	70× 5	14+ 7	6	2÷ 4	2
5 5	2	7	3	4	5− 6	1
140× 4	5	24× 2	6	6− 7	1	12× 3
7	5− 6	1	2	1− 3	13+ 5	4
1− 6	7	3− 4	1	2	3	5

117

3÷ 1	3	192× 4	6	30× 2	5	2− 7
84× 7	210× 6	3+ 1	2	4	3	5
4	7	2	90× 5	3	6	1
3	5	13+ 6	7	42× 1	16+ 4	2
10+ 5	4	15× 3	1	7	13+ 2	6
3÷ 2	1	5	3 3	6	7	4
6	5− 2	7	1− 4	5	1	3

118

5+ 2	35× 7	5	2− 4	6	3÷ 1	3
3	1− 5	6	3+ 1	2	14+ 7	4
168× 7	2− 4	2	30× 6	5	3	1 1
4	6	21+ 7	3÷ 3	1	1− 5	3− 2
3+ 1	2	3	7	4	6	5
14+ 5	3	3+ 1	2	3− 7	4	15+ 6
6	5+ 1	4	15× 5	3	2	7

119

72× 3	4	6− 1	7	60× 6	5	2
15× 1	6	4− 7	3	3+ 2	15+ 4	5
5	3	2÷ 2	2− 4	1	6	6− 7
12+ 2	17+ 7	4	6	8+ 5	3	1
4	5	3	2	6− 7	1	6 6
6	2÷ 2	1− 5	3÷ 1	3	168× 7	4
7 7	1	6	20× 5	4	2	3

120

6− 1	56× 2	7	60× 5	1− 3	4	3− 6
7	4	10+ 1	6	2	30× 5	3
3	1	5	48× 4	6− 7	6	5− 2
1− 6	5	4	3	1	3+ 2	7
10× 5	7 7	5+ 3	2	168× 6	1	1− 4
2	13+ 3	9+ 6	1	4	7	5
4	6	2	2− 7	5	3÷ 3	1

121

20× 4	2− 7	5	2÷ 3	6	2÷ 1	2
5	14+ 2	48× 4	6	21× 7	3	1
7	5	6− 1	2	19+ 4	6	3− 3
6+ 2	3− 3	7	15× 1	5	4	6
1	6	2 2	5	3	70× 7	16+ 4
3	72× 4	13+ 6	7	1	2	5
6	1	3	2÷ 4	2	5	7

122

28× 4	4− 1	5	2÷ 6	3	5− 2	7
7	2÷ 2	1	9+ 5	4	16+ 6	2− 3
5+ 1	4	2÷ 6	3	5− 2	7	5
2÷ 6	1− 5	4	2÷ 2	7	3	5+ 1
3	42× 7	2	1	6 6	9+ 5	4
1− 5	6	3	6− 7	1	4	3÷ 2
1− 2	3	3− 7	4	4− 5	1	6

123

4− 1	5	42× 3	7	13+ 2	6	168× 4
5+ 3	2− 4	2	1 1	5	7	6
2	6	1− 4	14+ 5	1 1	1− 3	4− 7
1− 4	3+ 1	5	6	13+ 7	2	3
5	2	15+ 7	3	6	1− 4	3+ 1
13+ 6	7	1	24× 4	3	5	2
7	2÷ 3	6	2	10+ 4	1	5

124

3÷ 6	2	2− 4	2− 3	5	6− 7	1
3÷ 3	1	6	42× 7	3− 2	5	84× 4
2÷ 2	2− 5	3+ 1	6	360× 4	3	7
4	7	2	1	3	6	5
11+ 7	4	3 3	8+ 5	1	2	9+ 6
6+ 5	2÷ 3	2− 7	2− 4	6	1	2
1	6	5	5− 2	7	1− 4	3

125

2− 1	168× 7	3	4	13+ 5	2	6
3	8× 1	2	13+ 7	15+ 6	4	2− 5
2	4	13+ 7	6	1 1	5	3
35× 7	10+ 5	6	2÷ 2	4	3− 3	3− 1
5	2	5+ 1	21× 3	7	6	4
24× 6	3	4	7+ 5	2	15+ 1	7
4	1− 6	5	2− 1	3	7	2 2

126

2÷ 2	4	28× 7	8+ 1	1− 6	2− 5	36× 3
14+ 6	1	4	7	5	3	2
7	8+ 2	5	24× 3	4	2− 6	1
12+ 5	2÷ 3	1	2	105× 7	4	6
4	6	4− 2	5	3	6− 1	7
3	7 7	6	3− 4	1	14× 2	9+ 5
6+ 1	5	11+ 3	6	2	7	4

127

2÷ 4	2	5 5	126× 6	3÷ 1	3	6− 7
2− 3	5+ 4	2− 6	7	5− 2	1− 5	1
5	1	4	3	7	6	2÷ 2
1− 7	6	12+ 2	30× 5	3	11+ 1	4
3+ 1	2− 3	7	2	6	4	1− 5
2	5	3	112× 1	4	7	6
6 6	6− 7	1	4	30× 5	2	3

128

112× 4	2− 1	3	15+ 7	5	17+ 6	3− 2
7	4	12+ 6	3	2 2	1	5
3+ 1	2	4	1− 5	2÷ 6	3	7
2− 5	7	2	6	3	3− 4	3÷ 1
11+ 6	30× 5	6+ 1	2÷ 2	4	7	3
3	6	5	6− 1	7	2÷ 2	4
2	4− 3	7	3− 4	1	1− 5	6

129

2÷ 2	1	3− 7	4	1− 3	2÷ 6	60× 5
5− 1	1− 5	4	4− 7	2	3	6
6	6− 7	1	3	2− 5	10+ 4	2
2÷ 4	2	5− 6	1	7	5	84× 3
4− 7	3	36× 2	7+ 5	24× 6	1	4
60× 5	6	3	2	4	7	1
3	4	1− 5	6	3+ 1	2	7 7

130

14+ 2	7	5− 1	2− 6	4	20× 5	1− 3
13+ 7	5	6	1− 3	2	1	4
6	12+ 2	4− 7	14+ 5	3÷ 3	4	1
4	6	3	2	1	2− 7	5
15× 3	3− 1	2÷ 4	7	35× 5	14+ 6	2
5	4	2	1	7	12+ 3	6
2− 1	3	20× 5	4	6 6	2	7

131

12× 1	2	6	12+ 3	4	5	3− 7
2− 5	2÷ 6	6− 7	1	1− 3	2	4
7	3	9+ 5	7+ 2	1	4	3÷ 6
24× 4	1	3	12+ 7	1− 5	6	2
6	1− 4	1	5	5− 2	7	3÷ 3
12+ 2	5	4 4	1− 6	7	15× 3	1
3	7	48× 2	4	6	1	5

132

3− 1	2÷ 4	2	4− 5	13+ 7	6	16+ 3
4	6× 2	3	1	1− 5	7	6
2÷ 6	1− 7	2÷ 1	2	4	8+ 3	5
3	6	3− 4	7	4− 1	5	2− 2
35× 7	2− 3	5	6 6	6× 2	1	4
5	1	6 6	1− 4	3	12+ 2	6− 7
70× 2	5	7	3	6	4	1

133

1− 3	2− 7	3÷ 2	1− 5	6	3− 4	2÷ 1
4	5	6	7 7	14+ 3	1	2
6+ 1	30× 2	5	3	7	72× 6	4
5	24× 6	1	10× 2	4	5− 7	3
4− 7	3	4	1	5	2	1− 6
3÷ 2	3− 4	13+ 7	6	1	15+ 3	5
6	1	3 3	2÷ 4	2	5	7

134

30× 6	5	1	13+ 2	4	7	12× 3
1	10× 2	5	2− 7	15+ 3	6	4
6− 7	1	3− 4	5	6	1− 3	2
9+ 5	2÷ 3	7	5+ 4	1− 1	3÷ 2	6
4	6	2÷ 3	1	2	13+ 5	7
2− 2	4	6	11+ 3	12+ 7	1	6+ 5
21× 3	7	2	6	5	4 4	1

135

1− 6	6− 7	2÷ 2	1	2− 3	5	12+ 4
5	1	18× 6	15+ 7	2	2÷ 4	3
42× 7	3	1	80× 4	6	2	5
2	1− 6	1− 3	5	4	6− 1	5− 7
3	5	4	14+ 6	1	7	2
2÷ 4	2	105× 5	3	7	3− 6	5− 1
3− 1	4	7	7+ 2	5	3	6

136

1− 6	10× 5	2	1− 4	3	8+ 7	1
7	1− 3	3− 5	2	11+ 1	6	4
3− 1	4	3÷ 3	13+ 5	6	2	18+ 7
4	14× 2	1	2− 7	5	1− 3	6
3 3	7	9+ 6	1	2	4	5
3− 2	5− 1	13+ 7	6	140× 4	5	1− 3
5	6	1− 4	3	7	1 1	2

137

3− 7	4	54× 6	3	1	4− 5	7+ 2
3÷ 6	2	13+ 7	4	3	1	5
3− 1	2÷ 6	3	2	7+ 5	11+ 4	7
4	11+ 3	4− 1	1− 5	2	1− 7	6
2− 3	7	5	6	24× 4	2	12× 1
5	1	2 2	1− 7	6	3	4
3− 2	5	3− 4	1	1− 7	6	3

138

30× 5	6	4 4	12+ 2	6− 7	1	4− 3
1	6× 2	3	4	1− 5	6	7
3÷ 2	1	2− 5	6	84× 3	7	4
6	3− 4	7	6+ 1	2	3	10+ 5
36× 3	7	3− 2	5	30× 6	4	1
4	3	17+ 6	7	1	5	3÷ 2
2− 7	5	1	3	2÷ 4	2	6

139

3− 1	3÷ 2	6	2− 7	60× 5	3	4
4	3÷ 3	1	5	12+ 7	2	4− 6
30× 6	5	6− 7	1	3	3− 4	2
3÷ 2	6	1− 5	4	11+ 1	7	3÷ 3
12+ 5	3− 7	3 3	7+ 2	4	6	1
7	4	2	3	12+ 6	1	5
2− 3	1	12+ 4	6	2	2− 5	7

140

2÷ 4	2÷ 3	490× 2	7	5− 1	6	13+ 5
2	6	7	5	3	4	1
11+ 7	9+ 2	3÷ 1	3	15+ 4	210× 5	6
1	4	3	6	5	7	2 2
3	1 1	10+ 5	4	3÷ 6	1− 2	3− 7
11+ 5	1− 7	6	1	2	3	4
6	20× 5	4	5− 2	7	3÷ 1	3

141

1− 2	3	120× 6	4	5	6− 1	7
360× 6	4	6+ 2	1	3	12+ 7	5
3	13+ 6	7	2÷ 2	1	1− 5	4
5	3− 1	4	17+ 7	6 6	24× 3	3÷ 2
6− 1	3− 2	4− 5	3	7	4	6
7	5	1	1− 6	12+ 4	2	2− 3
3− 4	7	3 3	5	2	6	1

142

360× 4	6	5	3	2÷ 2	6− 1	7
1− 3	2	2− 7	5	4	3− 6	2÷ 1
12+ 5	7	6× 1	3− 4	1− 6	3	2
3− 2	5	6	1	7	12× 4	3
6+ 1	3	2	1− 6	5	16+ 7	4
1− 6	3− 4	12× 3	8+ 7	1	2 2	5
7	1	4	1− 2	3	1− 5	6

143

80× 4	17+ 7	3	11+ 5	6	12× 1	2 2
5	4	7	1− 3	2	6	3÷ 1
13+ 1	6	5	4	18+ 7	2	3
8+ 3	1	3÷ 6	2	5	4	2− 7
2	3	2− 4	6	2− 1	2− 7	5
1− 6	2÷ 2	1	6− 7	3	5	10+ 4
7	3− 5	2	1	1− 4	3	6

144

14+ 6	2÷ 1	2	1− 4	5	13+ 3	7
5	24× 6	4	10+ 7	3	2	1
3	18+ 4	21+ 7	10+ 1	2	11+ 5	6
4	3	6	2 2	7	24× 1	7+ 5
6− 1	7	3	5	4	6	2
7	8+ 2	5	15+ 6	28× 1	4	1− 3
3− 2	5	1	3	6	7	4

145

16+ 5	3	4− 2	24× 4	6	1− 1	28× 7
7	1	6	105× 5	3	2	4
6+ 2	12+ 4	2− 5	3	7	7+ 6	1
3	7	1	11+ 2	5	84× 4	13+ 6
1	1− 2	3	1− 6	4	7	5
144× 6	1− 5	4	7	2÷ 1	3	2
4	6	8+ 7	1	2	8+ 5	3

146

9+ 1	4	6× 3	3− 2	5	1− 7	6
4	420× 6	2	6− 7	1	11+ 3	4− 5
2÷ 3	5	7 7	3− 4	2	6	1
6	2	1− 5	1	8+ 3	4	4− 7
14+ 2	7	4	11+ 5	2− 6	1	3
7	5+ 3	1	6	4	11+ 5	2
5	1	2÷ 6	3	14× 7	2	4

147

3− 1	18× 6	1− 2	3	18+ 7	4	10× 5
4	1	3	60× 6	5	7	2
18+ 7	40× 5	1	4	2	2÷ 6	3
6	2	16+ 5	6− 7	5− 1	2− 3	4 4
2	3	4	1	6	5	1− 7
19+ 5	4	7	6+ 2	3	1	6
3	7	30× 6	5	8× 4	2	1

148

1− 2	15× 5	3	1 1	10+ 6	4	35× 7
3	6− 1	72× 4	6	5− 2	7	5
16+ 4	7	18+ 5	3	12× 1	6	2
5	6	7	7+ 2	3− 4	3÷ 1	1− 3
1	14+ 2	6	5	7	3	4
6	4	2	11+ 7	2− 3	5	5− 1
11+ 7	3	1	4	3− 5	2	6

149

60× 5	6	2− 3	2÷ 2	1	14+ 7	4 4
2	6− 1	5	1− 7	6	4	3÷ 3
120× 6	7	9+ 4	5	9+ 2	3	1
4	5	1− 7	6	3	6+ 1	9+ 2
4− 3	6× 2	5− 6	1	4	5	7
7	3	10+ 1	4	3− 5	2	30× 6
3− 1	4	2	3	13+ 7	6	5

150

1− 5	4	210× 7	6+ 2	1	3	24× 6
6	5	1	3÷ 3	4− 2	14+ 7	4
12+ 2	7	1− 5	1	6	4	3
7+ 1	3	6	20× 4	5	2 2	210× 7
3	1	2	3− 7	4	6	5
11+ 7	4− 2	2− 4	6	4− 3	6+ 5	1
4	6	2− 3	5	7	2÷ 1	2

Made in the USA
Monee, IL
13 September 2026

62976623R00079